BRITISH
STEAM ENGINES

igloo

LNER 'A4 Pacific' class 4–6–2 No. 4468 'Mallard', built in 1938 to a design by Nigel Gresley, which holds the world speed record for steam locomotives at 125.88 mph. Here she hauls a steam special over the viaduct at Knaresborough, Yorkshire, in April 1987. 'Mallard' is housed permanently at the National Railway Museum.

BRITISH STEAM ENGINES

INTRODUCTION BY O.S. NOCK

Published in 2011
by Igloo Books Ltd
Cottage Farm
Sywell
NN6 0BJ
www.igloo-books.com

ISBN 978-0-85734-802-9

10 9 8 7 6 5 4 3 2 1
M044 0311

Author team: Jon Mountfort, Tom Dodds,
Tony Evans, David Adams

Printed and manufactured in China

LMS Princess Coronation class 'Duchess of Sutherland'
No. 6233 at Morton, Lincolnshire, hauling the 'Yorkshire
Coronation' on a steam special.

CONTENTS

ACKNOWLEDGEMENTS

Cover images: Front cover top images from left to right: Szczepko/Dreamstime, Georgesixth/Dreamstime. Front cover bottom images from left to right: Chris Jenner/Shutterstock, Colin Hutchings/Shutterstock. Front cover main image: Louisep/Warren Palmer's first collection/fotolibra. Most of the images in this book come from the 'Memories of Times Past' archive of historical illustration. In addition we would like to thank the following for the use of paintings and photographs: Ivan Abrams (185); Jon Bennett (154); Peter Brabham (81, 211); Matt Buck (151B); Roger Carpenter (138B); Les Chatfield (18, 58T, 119B, 164, 191, 212–13); Geoff Cryer (203); The Terence Cuneo Trust (133); Peter Ellison (200-201); Alan Fearnley (6–7); Brian Forbes (152–3); Barry Freeman (46–7, 77, 102–3, 128–9); Al Green (218); Paul Gribble (11); Dave Hamster (94MR); Brian Harrington Spier (118B); Duncan Harris (139, 167, 172, 215); Andrew Henwood (112B); S. C. Hine (97); Tony Hisgett (15, 68-69, 107T, 107B, 142, 166, 173, 189, 194, 210, 217); June Hoyle (1); David Ingham (60, 62, 116, 121, 141BL, 146B, 147B, 162–3, 165, 175, 206); David Jones (158); Tim Jones (195); Roy Lambeth (63BR); J. Lord (159); Stan Marston (140, 190); Nick McLean (57); Wendy Meadway (9); David Merrett (216); Charles Miller (156TL); Brian Negus (72B); Matt Ots (58BL); Richard Outram (176–7); Phil Parker (183); Rob Phillips (178); Richard Ratcliffe (207); Malcolm Root (64–5, 156–7); Ben Salter (145); Phil Sangwell (4–5, 171, 204); Ray Schofield (127) Colin Smith (2-3, 123); Simon Tegg (16–17); Rob Terrace (150); Paul Tomlin (151 TL); Stella Whatley (125); Clare Wilkinson (174); Ian Willcox (160); Graham Williams (78B); David S. Wilson (205); London Science Museum (28B); Swindon Steam Museum (41T); Wikimedia Commons (16T, 58M, 61TR, 73TR, 95ML, 124, 137T, 138TR, 181, 182, 188, 198); and Zabdiel (161).

INTRODUCTION BY O.S. NOCK

Oswald Stevens Nock, 'Ossie' to his many friends and colleagues, is probably the world's most prolific and well-respected author on the subject of railways – when he died in 1994 he had just completed his 143rd railway book.

He was born in the West Country in 1905. Having graduated in engineering from Imperial College, London, Nock joined the staff of Westinghouse Brake and Signal Company in 1925, with whom he remained for 45 years, holding many senior positions, including Chief Mechanical Engineer and Planning Manager. 'I always thought of writing as a second string to the bow,' he said. 'The real job was as an engineer.'

Ossie met his wife, Olivia, at King's Cross railway station, where she was assistant manageress in the Georgian Tea Rooms. His other real love was his model railway. So large was his layout that, when the Nocks moved to a newly-built bungalow near Bath in 1980, they needed two houses – one for the family and one for Ossie's model railway.

The introduction to this book is an expanded version of the introduction to one of Nock's best-selling titles, British Steam Locomotives, first published by Blandford Press in 1964. It has been brought up to date by Jon Mountfort.

INTRODUCTION

Railways, in a number of crude forms, had existed for many years before the birth of George Stephenson. By putting the primitive wagons of the day on to rails, instead of trundling them along rough and badly maintained roads, it was found that horses could pull far heavier loads and the Stockton and Darlington Railway, which is generally considered to be the springboard from which the railway system of Britain and the world originated, was laid out for horse traction. The geography of County Durham aided this general design. The country falls from the coalfields of Bishop Auckland to the sea and horses could comfortably manage long trains of heavily loaded wagons and equally well haul the empties back to the colliery districts.

At the opening of the line in 1825, the Company had just one steam locomotive, the famous 'Locomotion No. 1', that is now exhibited in the Darlington Railway Centre and Museum and in the great vision of development held by George Stephenson, she was to be the fore runner of many more.

The original 'Rocket' of 1829, showing cylinders mounted at a steeply-inclined angle high up on the side of the boiler which gave her an awkward swaying motion as she ran.

For many years, George Stephenson and his supporters had to fight an uphill battle against those who still favoured horse traction. The issue was settled beyond doubt by the trials at Rainhill on the line of the Liverpool and Manchester Railway in October 1829, when the engine built by George Stephenson's son, Robert, won the prize offered by the directors for the best locomotive that could fulfil certain speed and haulage conditions. This locomotive was the 'Rocket', which was the true progenitor of the machine that was to revolutionize inland transport in Great Britain. The Liverpool and Manchester Railway was the first main-line railway between two cities in the world. Its backers envisaged that it would generate most of its income from the transport of minerals and the products of the new industries that were growing up in the country. Once it opened for business, they were pleasantly surprised by the success of its passenger operations.

The steam locomotive not only provided means for enabling people to travel faster, but it made travel cheap. Simple souls, who had never ventured beyond the confines of their own village, could now travel at a penny a mile, or even less, on special excursion trips. Railways made the flow of commerce easier; merchants could travel more easily and faster to see their clients; the day of the 'commercial traveller' was at hand.

But perhaps of greater benefit to the country as a whole was the way in which railways, through the haulage capacity of the steam locomotive, brought down the price of consumer goods. The opening of the Great Northern Railway, from Doncaster to London, in 1852, showed what could be done. The result of its enterprise and the effectiveness of the powerful locomotives used, was to revolutionize transport that the price of household coal in London was reduced from 30 shillings a ton to 7 shillings.

Although steam traction made travel cheap and easy, for the masses it was relatively uncomfortable. While the rich travelled first class in enclosed carriages, humbler folk experienced the pleasure of seeing something of the country from the open, third-class 'carriages' which were little more than coal wagons fitted with bench seats. Harrowing pictures have been painted of the miseries of travelling 'third' in those days and the conditions certainly could be harsh in wet or wintry weather; but early excursion trains, however slow and draughty, were very popular and undoubtedly contributed to the gradual spread of the broader education of the nineteenth century.

A contemporary lithograph of the opening of the Stockton and Darlington Railway in 1825.

The success of steam locomotives, in both goods and passenger services, fostered the development of new engineering processes: in the iron and brass foundries, in heavy forging and in the manufacture of iron itself. Railway needs led to improvements in technology that came to benefit the trade of the country as a whole. The need for something harder and stronger than wrought iron for wheels, rails and other moving parts subject to hard wear, fostered the rapid development of steel. The very first plant anywhere in the world for manufacture of steel on a commercial scale was at the Crewe locomotive works of the London and North Western Railway in 1864. It was extensively used, not only for the production of locomotive parts, but to roll the rails they ran upon.

One is entitled to look with some awe and reverence to the unique, self-contained piece of machinery that was indeed the phenomenon of the nineteenth century. From the dawn of history, the fastest mode of transport known to man was that of a galloping horse, and yet, within twenty years of the opening of the Stockton and Darlington Railway, there were steam locomotives capable of travelling at 95 km (60 miles) per hour. By the 1870s, speeds of 130 km (80 miles) per hour were occasionally being touched and by 1904 a speed of 160 km (100 miles) per hour had been attained by 'City of Truro' on the Great Western Railway.

The pioneering LNER locomotive works at Crewe.

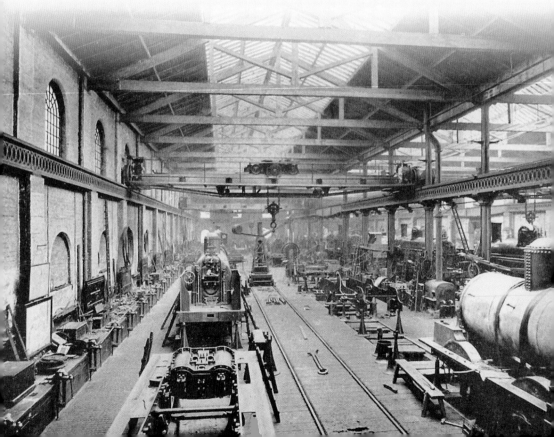

Inevitably, the locomotive became the centrepiece of the entire railway scene. The passengers might be carried in open trucks and later in roofed affairs that strongly suggested a cavalcade of horse-boxes, but the locomotives were arrayed in a splendour of polished metal and gleaming paint and their drivers and firemen delighted in keeping them spotless. Among the larger railways, the heights of magnificence were reached in the last years of the 19th century and, on such lines as the Great Western, the Midland, the Brighton, the North Eastern and the Highland, the dazzling array could not have been more ornate, yet in excellent taste.

From 1880 onwards, engineers began to seek means of securing greater efficiency in operation and of reducing the coal bill. In marine engineering, double- and triple-expansion engines were, by that time, commonplace and many trials were made with steam locomotives having compound, or two-stage expansion of the steam. None of the nineteenth-century experiments in Great Britain had any lasting success – chiefly because compounding brought with it certain gadgets, the ineffectiveness of which far outweighed any increase in thermal efficiency that might have been secured by compound rather than single-stage expansion. It was the purchase of a French-built de Glehn compound, 'La France', in 1903, by the Great Western, that virtually dealt the death-blow to the compound principle in Great Britain for at Swindon, George Jackson Churchward, one of the greatest masters of steam locomotive design, built single-expansion machines that were the superior of the French compounds.

GWR locomotive 'La France' at Bristol Temple Meads station in 1905, in a painting by Paul Gribble.

Railways, as required by British law, were expensive things to build and run and long before the days of motorcars and lorries, something less pretentious was felt to be necessary. The legislation, providing for numerous relaxations from main line standards, was given by the Light Railway Act of 1896 and a most picturesque line built under these provisions was to be seen in the Leek and Manifold Railway in the Derbyshire hills. These narrow gauge lines, like the Lynton and Barnstaple in Devon, provided many attractions for the railway connoisseur, but few of them made any money for their owners, the most famous of whom was Colonel Stephens, who was involved with the building and running of more than fifteen light railways in the early 1900s.

In the twentieth century, on the main lines, the form and styling of locomotives is in some instances a reflection of national affairs. One by one, the exuberant liveries of old began to disappear. Goods engines in ever-increasing numbers began to be painted black and the onset of the

Great War soon witnessed the painting over of much of the polished brass and copper work. The South-Eastern and Chatham, the Great Northern and the Great Eastern, all renowned for the gorgeous turnout of their express locomotives, changed to various shades of dull grey. These were signs of the times and some of the old liveries were gone, never to reappear. Then came the grouping of all the old independent railways of Great Britain into four large companies.

From 1923 onwards, the kaleidoscope of locomotive liveries that had so distinguished the railways of this country was suddenly contracted to four styles, of which three were green – those of the Great Western, of the Southern and of the London and North Eastern. The London Midland and Scottish Railway retained 'Derby-red', but very soon all, except the top-line passenger types, were painted black. The Somerset and Dorset Joint and the Midland and Great Northern Joint retained their distinctive colours and there were some bright interludes, as when Sir Nigel Gresley built the Silver Jubilee train in silver throughout – engine and coaches alike – and when the streamlined 'Coronation' trains of 1937 onwards were finished in blue or maroon.

L&M narrow gauge and NSR standard gauge locomotives on opposite sides of the platform at Waterhouses Station, Leek and Manifold Railway, around 1920.

The 1930s were a time of great financial anxiety for the railways of Great Britain and with traffic dwindling, many devices of the showman's art were tried to win business. The streamlining of locomotives had a value far greater to the publicists than to the engineers. At the same time, speeds were soaring. For the first time, a British train had made a considerably long run at an average speed of more than 130 km (80 miles) per hour from start to stop. The maximum speed record went up from 160 km (100 miles) per hour, made by the Great Western 'City of Truro' in 1904, to 174 km (108 miles) per hour by the Gresley non-streamlined Pacific 'Papyrus' in 1935; to 181 km (112.5 miles) per hour by the first LNER streamliner 'Silver Link' later that same year, and then to 183 km (114 miles) per hour by the Stanier 'Pacific Coronation' in 1937. Finally, in 1938, there came the record that is likely to stand for all time – the British and World record for steam of 203 km (126 miles) per hour by the Gresley A4 streamliner, 'Mallard'.

London and NER 4–6–0
locomotive, painted and lettered
in new style, July 1923

GWR down West of England
Express, 4-cylinder 4–6–0
locomotive No. 4099
'Kilgerran Castle', July 1927

Southern Railway 4–6–0 mixed
traffic locomotive No. 475E,
September 1924

LM&SR four-cylinder 4–6–0
locomotive, Western Division,
painted and lettered in new style,
January 1924

World War II brought an immediate end to all this activity. The locomotives built for war service include the austerity 2–8–0, the LNER B1 and the LMS 8F 2–8–0, which was ordered to be built for general service by a number of the once-independent railway works. The war was hardly over when nationalization of the railways took place and the experience of the four main-line companies was pooled in the work of a newly formed Railway Executive. From the plethora of strongly individual practices, a single set of locomotive designs for the whole country was evolved by Robert Riddles and his team and although naturally there were some compromises and mistakes, the British Railways standard locomotives formed a worthy climax to the story. The last of all, the big 9F 2–10–0 freight engine, was in every way a remarkable machine: designed from the outset for hauling freight, it was just as happy heading fast express passenger trains.

The line of development, extending from Richard Trevithick's Pen-y-darren locomotive of 1804, through 'Locomotion No. 1' of 1825 and 'Rocket' of 1829 to the 9F 'Evening Star' of 1960, forms a deeply impressive record of a great national achievement. When steam was abandoned on British Railways in 1968, it was indeed the end of an era and one might have thought that steam engines would never be seen again on British main-line tracks. This is exactly what the authorities intended, for British Rail management introduced a ban on steam on the public railway from this date forward and preserved engines were confined to chugging slowly up and down short stretches of track, which had been rescued from old closed lines. But enthusiasts for steam, of which there are countless thousands in Britain, never give up without a fight and by campaigning vigorously against the ban, BR relented; in 1972 it was lifted. Since they were allowed back on the network, preserved steam locomotives have regularly been seen, once more, puffing along in their natural setting.

Then, in 2008, something amazing happened: the first brand-new main-line steam locomotive turned its wheels on British metals since 'Evening Star' emerged from the workshops in Swindon in 1960. 'Tornado' is a copy of a Peppercorn A1 Pacific, of which there are no surviving originals. It was conceived by a group of dedicated enthusiasts around 1990 and was paid for by small monthly subscriptions raised from individuals. It made its first run on the public rail network on the evening of 4 November 2008 between York and Scarborough and now makes regular journeys carrying passenger excursions all over Britain. Steam on British railways lives on.

'Coronation Scot' on her way to the New York World Fair, July 1935 (left)

'Tornado' at Birmingham (right); and en route to Ropley on her maiden journey (below).

HOW A STEAM LOCOMOTIVE WORKS

All steam locomotives essentially consist of three parts – the firebox, the boiler and the cylinders.

The firebox contains a large coal fire. Good quality coal provides a hot, evenly-distributed fire. Producing a good fire was a skilled and heavy job, the fireman having to shovel tons of coal over a long journey to power a main-line locomotive.

The smoke and hot gases from the fire pass through a hundred or more small diameter tubes in the body of the boiler. These create a very large surface area from which to heat the water to produce steam, which is collected in a dome at the top of the engine. To improve efficiency, the steam is then superheated by passing it along pipes through more fire tubes in the boiler. A safety valve fitted to the boiler prevents the steam pressure from getting too high and the engine exploding.

The driver uses a regulator to control the flow of steam to the cylinders. Valves allow steam to be injected into each side of the cylinder in turn, so that each stroke of the piston is a power stroke. Exhaust steam is passed from the cylinder through a blastpipe to the chimney, causing a partial vacuum. The vacuum draws the hot gases from the fire through the boiler tubes – the faster the engine runs, the faster the heat is drawn through the boiler and the more steam is produced. The release of exhaust steam causes the familiar 'chuff' sound.

The piston in the cylinder is connected to the driving wheels through connecting rods and cranks. Most engines use two cylinders – one on each side of the engine. The cranks on either side are offset by 90 degrees, thereby evenly spreading the power from the two double-acting cylinders during a complete revolution of the wheels and enabling the engine to start from rest, whatever the position of the pistons in the cylinders.

cylinder

The major components of a piston steam engine, typical in a steam locomotive. This is called a 'double-acting steam engine', because the valve allows high-pressure steam (shown in red) to act alternately on both faces of the piston. The blue indicates exhaust steam.

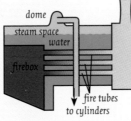

dome

steam space

water

firebox

fire tubes

to cylinders

A simplified cross section of a locomotive boiler, showing how the smoke tubes heated from the firebox heat water to create steam, which is then forced out to drive the cylinders.

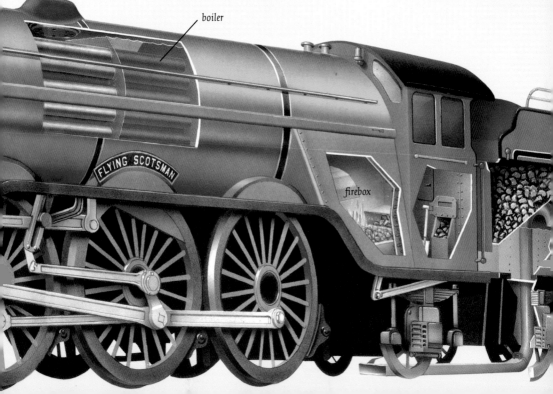

boiler

FLYING SCOTSMAN

firebox

STEAM LOCOMOTIVE WHEEL ARRANGEMENTS

The Whyte system is commonly used to describe steam locomotive wheel arrangements in Great Britain and North America. From left to right, separated by dashes, it gives the number of leading wheels, the number of driving wheels and the number of trailing wheels of the locomotive. The table opposite shows the wheel arrangements along with their nicknames, which are mostly of American origin.

The driving wheels of 'Southern Maid', a 4–6–2 Pacific locomotive, designed by Henry Greenly and built by Davey Paxman & Co. in 1926 for the Romney, Hythe and Dymchurch Railway.

TANK ENGINE DESIGNATIONS

The Whyte system sometimes has letters after the wheel arrangement. Common examples are 0–6–0T or 2–6–4T where the T denotes that this is a 'tank engine' as opposed to a 'tender engine'. In a tank engine, the water and coal are both carried on the locomotive, the advantage being that the engine can run backwards for a whole journey, rather than having to be turned round to face the right way using a turntable. The position of the water tank on the locomotive is also sometimes specified by letters. For example, 0–4–0ST is a 'saddle tank' in which the tank is wrapped around the top of the boiler, like a saddle. 0–4–0PT is a 'pannier tank' in which there are two separate tanks, one on each side of the boiler; the tanks don't extend all the way down to the frames, so they look like panniers. 0–4–0WT is a 'well tank' in which the tank is beneath the boiler, between the frames. If the letter T is on its own, this usually denotes a 'side tank' locomotive, which is like a pannier tank, but with the tanks going all the way down to the frames.

UNUSUAL WHEEL ARRANGEMENTS

The Whyte system can also be used to describe some of the less common steam locomotive wheel arrangements, such as the Fairlie, the Mallet and the Garratt. These locomotives have two sets of driving wheels, which are

Leading-Driving-Trailing	Arrangement	Name
	0-2-2	
	2-2-0	
	2-2-2	Patentee
	4-2-2	
	0-4-0	
	0-4-2	
	2-4-0	
	4-4-0	American
	4-4-2	Atlantic
	0-6-0	Six-coupled
	0-6-2	
	2-6-0	Mogul
	2-6-2	Prairie
	2-6-4	
	4-6-2	Pacific
	4-6-4	Hudson
	0-8-0	Eight-coupled
	2-8-0	
	2-8-2	Mikado
	0-10-0	Decapod
	2-10-0	
	2-4-4-2	Mallet
	2-6-6-2	Mallet
	0-4-0+0-4-0	Fairlie/Garratt
	0-6-0+0-6-0	Fairlie/Garratt
	4-4-2+2-4-4	Garratt
	2-6-2+2-6-2	Garratt

independent of one another, so extra numbers are used to denote the two sets. Note that the following table only gives a few examples of these layouts. Some, such as the American 4–8–8–4 and 2–6–6–6 Mallets, were absolutely enormous – the largest and most powerful steam locomotives ever built. The 59 class Garratts, operated by East African Railways, were of 4–8–2+2–8–4 wheel arrangement.

View of the railway across Chat Moss, on the Liverpool and
Manchester Railway, 1831, a painting by Henry Pyall based on an
engraving by Thomas Talbot Bury.

THE DAWN OF THE STEAM LOCOMOTIVE, 1800–1850

During the first decade of the nineteenth century, entrepreneurial spirit and advances in engineering transformed the static steam engines developed in the preceding century, for various industrial purposes, into the first steam locomotives. The most significant of these early experiments was Richard Trevithick's high-pressure steam engine, driven along an existing tramway at a South Wales ironworks in 1804. Further technical improvements over the next twenty years saw the use of steam locomotives become widespread in industrial premises, inspiring other engineers, including George Stephenson, his son Robert, Isambard Kingdom Brunel and Daniel Gooch, to turn their attention to the potential of steam railways. *The introduction to this book is an expanded version of the introduction to one of Nock's best-selling titles, British Steam Locomotives, first published by Blandford Press in 1964. It has been brought up to date by Jon Mountfort.*

THE DAWN OF THE STEAM LOCOMOTIVE, 1800–1850

The nineteenth century was the century of steam power. In little more than twenty-five years, technological improvements saw the steam engine transformed from large, underpowered and static monsters into powerful, mobile locomotives not dissimilar to those that dominated Britain's railways until the 1960s.

However, the use of steam to produce mechanical force was not a new concept. Almost two thousand years ago, Hero of Alexandria described how steam escaping from nozzles caused a sphere to rotate. In the seventeenth century, John Wilkins, the first secretary of the Royal Society, described a device using steam to rotate a roasting spit.

A number of inventors turned their attention to using steam power for industrial applications. In 1698, the English inventor, Thomas Savery, patented his 'fire-engine'

as a means of pumping water from mines. These early engines used low-pressure steam let into a vessel. Condensing the steam induced a partial vacuum and the atmosphere, acting against the vacuum, forced water from the mine and up into the vessel. Savery's engines had no moving mechanical parts apart from the taps which allowed steam to enter the vessel, water to be added for cooling and the condensed water to run off. Because these engines relied on atmospheric pressure acting directly on the water, they could not work below a depth of 9 m (30 ft) and were a failure in every instance where they were tried.

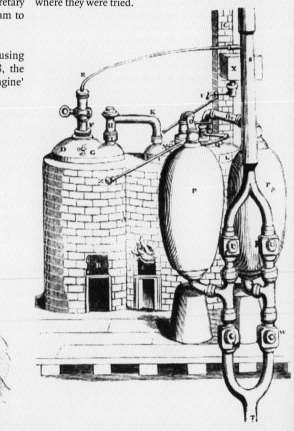

Thomas Savery,
and his steam
pump of 1698.

NEWCOMEN'S ENGINES

Invented by Thomas Newcomen, the 'atmospheric engine' built in 1712 used a piston in a cylinder and is generally regarded as the first practical steam engine. Despite the fact that Savery's engines never successfully drained a mine, his patent covered all engines that raised water using a fire. Consequently, Newcomen was forced to go into partnership with Savery. Newcomen's engine used low-pressure steam to fill the cylinder. Valves shut off the steam and introduced a spray of water to condense the steam in the cylinder, producing a partial vacuum beneath the piston. Atmospheric pressure above the piston pushed it down, giving a power stroke that acted on a rocking beam to operate a water pump, which could be 30 m (100 ft) or more below. Releasing the vacuum allowed the weight of the pump to drag the piston back to the top of the cylinder and the next cycle would commence by letting in fresh steam. Newcomen's engine was an immediate success and was used across Great Britain to pump out flooded tin and coal mines.

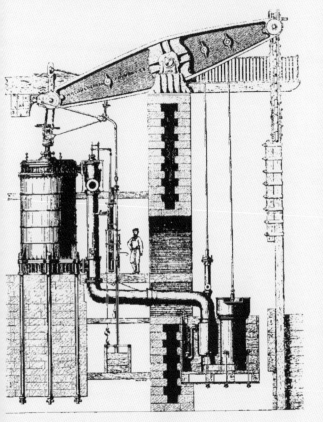

Scottish engineer, James Watt, patented a significant improvement to Newcomen's engine in 1769. Instead of cooling the steam in the cylinder, Watt provided a separate condenser connected by a pipe. When a valve on the pipe was opened, the vacuum in the condenser would act upon the cylinder below the piston, thus eliminating the wasteful cooling of the main cylinder. This also paved the way for the development of the double-acting engine, in which both upwards and downwards movements of the piston were power strokes.

Newcomen and Watt were both advocates of low-pressure steam. The boilers they used were of cast iron, the parts riveted together; a safety valve hadn't yet been invented. Boiler technology needed further development before high-pressure steam could be used successfully.

A Watt or Newcomen engine used atmospheric pressure of approximately 15 pounds per square inch (psi) to push the piston in its cylinder. The more power that was required, the bigger the piston area had to be and these early engines were huge.

A Newcomen engine installed in a tin mine in Cornwall, mid-eighteenth century.

TREVITHICK'S STEAM LOCOMOTIVES

Trevithick patented his high-pressure steam engine in 1802. To demonstrate the practicalities, a Trevithick stationary engine was installed at Coalbrookdale, Shropshire, working at an unprecedented 145 psi and producing 40 piston strokes a minute – a vast improvement on the 12 cycles a minute produced by a Newcomen and Watt engine.

It is thought that the Coalbrookdale Company then built a rail-mounted locomotive for Trevithick. In a letter dated 1802, Trevithick wrote: 'The Dale Co. have began a carriage at their own cost for the real-roads and is forcing it with all expedition'. Unfortunately, no other details of this experimental engine, or its success, or otherwise have survived. There is a drawing of a 'tram engine' from 1803, but it is not clear who designed it, or if it ever ran.

In 1803, Trevithick demonstrated another steam-powered road carriage. 'The London Steam Carriage' ran from Holborn to Paddington and back, but the poor state of the roads meant it was difficult to steer and uncomfortable to ride.

The following year saw Trevithick's first really successful locomotive. He had provided some of his high-pressure stationary engines to the Pen-y-darren ironworks at Merthyr Tydfil, in Wales. One was mounted on wheels and became the first true railway locomotive, moving a load of 10 tons of iron a distance of almost 16 km (10 miles) along the Company's tramway to Abercynon and returning with the empty wagons, in doing so winning a wager for 500 guineas made with Richard Crawshay. Although successful, the weight of the engine broke many of the tramway's cast-iron plate rails and it became stationary again. But this locomotive introduced a device that was fundamental to all steam railway engines for the rest of their history: Trevithick exhausted the spent steam through an upturned nozzle in the chimney, thus using the force of the steam to draw the fire. It was the first example of what was to become known as the 'blastpipe'.

The next Trevithick locomotive was built at Gateshead, near Newcastle upon Tyne, for the Wylam colliery wagonway (see page 27). Drawings of this engine survive. The relatively compact boiler provided steam for just one cylinder, which worked horizontally to drive a large flywheel. Drive was taken to the wheels by gears.

Trevithick's final foray into steam locomotives was to be the 'Catch Me Who Can', which ran on a circular demonstration track in London in 1808. Intended to show that locomotives could travel faster than horses, it failed to capture the public imagination and when the locomotive derailed, Trevithick had the attraction dismantled. Surviving sketches indicate that the engine used a vertical piston driving the rear wheels through a connecting rod, dispensing with the need for gears and a flywheel.

Trevithick's 1803 steam road carriage (left). A reconstruction of the Pen-y-darren locomotive at York Railway Museum (right).

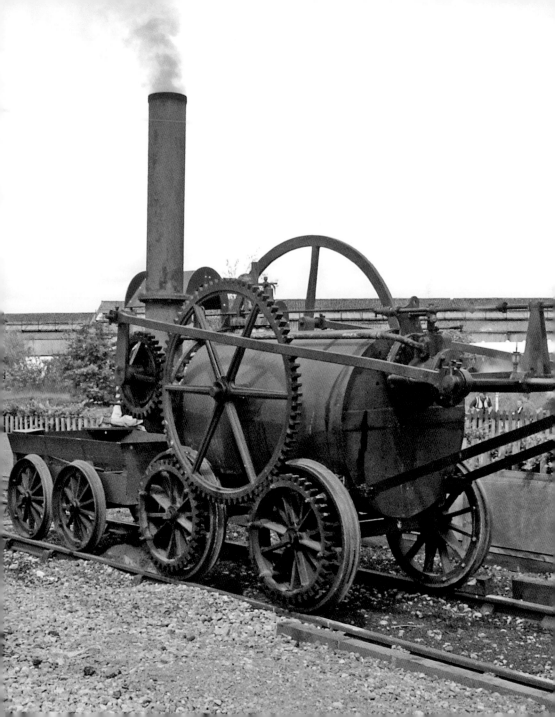

EARLY HIGH-PRESSURE STEAM ENGINES

By the start of the nineteenth century, stationary steam engines were commonplace in mines and factories. However, early engines built by Thomas Newcomen and James Watt used low pressure steam and were underpowered, large and far from mobile.

Richard Trevithick (1771–1833) saw that improvements in boiler technology made high pressure engines a practicality. He began work on engines using 'strong steam' at up to 50 pounds per square inch (psi), increasing the power to weight ratio of the engine and dispensing with the cumbersome condensers and boilers of Newcomen and Watt designs. Portable engines became possible.

In 1801, Trevithick demonstrated a steam-engined road carriage, but it caught fire when it was left unattended.

Undeterred, in 1802 he patented his high-pressure steam engine and demonstrated another steam-powered carriage, but it was difficult to control on uneven roads.

The logical step was to try the engine on a wagonway, where rails would remove the need to steer and provide a better surface. Trevithick had supplied some of his high-pressure stationary engines to the Pen-y-darren ironworks at Merthyr Tydfil, Wales. In the spring of 1804, one of these was mounted on wheels and placed on the wagonway, where it moved 10 tons of iron and returned with the empty wagons. The locomotive had a single boiler with a return flue mounted on a four-wheel frame. The piston was coupled via a crank to a large flywheel mounted on one side and connected to the driving wheels using cogs.

An early Trevithick-design locomotive working at Wylam Colliery, Northumberland, in the 1840s, painted by Thomas Hair (below).

While the trial proved that a locomotive could pull wagons using plain wheels on plain rails, many sections of the brittle cast-iron track were smashed by the engine's weight, made worse by the fact that it had no springs to cushion the load.

Trevithick persevered and in 1808, he demonstrated his new design, 'Catch Me Who Can', on a circular track somewhere near today's Euston Square in London. It was opened as a visitor attraction costing one shilling for a ride, but when the engine derailed, he gave up and never built another.

Trevithick's first true locomotive of 1803 (below), from J.G. Pangborn's classic 1894 work, The World's Rail Way.

Richard Trevithick painted in 1803, the year he built the world's first true steam locomotive (right)

The 1808 demonstration of Trevithick's new design in Euston Square, London (below right)

SMOOTH WHEELS ON SMOOTH RAILS

Although Trevithick had shown that smooth iron wheels would adhere sufficiently to smooth iron rails, the science was not fully understood. In 1811, John Blenkinsop, another Tyneside engineer, patented a system where an engine drove a cogged wheel along a toothed track. The locomotive, 'The Prince Regent', built by Matthew Murray of Leeds, used not one, but two cylinders operating vertically with four power strokes. The cylinders used sliding bars and connecting rods to convey power to the wheels. Two engines were built by 1812 and put to work at Middleton Colliery near Leeds and later near Newcastle-upon-Tyne.

At about the same time, the Wylam wagonway had been relaid using iron plate rails to 1500 mm (5 ft) gauge, rather than wooden edge rails. Mine owner Christopher Blackett was keen to clarify the issue of adhesion – just how well would smooth wheels interact with smooth rails before slipping? William Hedley, the colliery manager, Timothy Hackworth, colliery blacksmith and enginewright Jonathan Forster built a four-wheeled, hand-driven device and tested it pulling weighted wagons on a section of the wagonway. Hedley then added a steam engine to the contraption, but it didn't work very well, probably because of its single-flue boiler.

Once the adhesion limits had been proven, Hedley, Hackworth and Forster built 'Puffing Billy' between 1813 and 1814. Mounted on four wheels, the locomotive was powered by a boiler of 1.2 m (4 ft) in diameter and 3 m (10 ft) in length, which had a return flue, so the chimney and firebox were at the same end, and a blastpipe – both ideas of Trevithick. Two vertical cylinders towards the rear drove the wheels through levers and connecting rods to a central jackshaft geared to the two axles. Weighing 8 tons, the engine proved too heavy for the cast-iron plate rails. It was rebuilt on four axles to spread the weight, subsequently returning to four wheels when the L-shaped cast-iron plate rails were replaced by wrought-iron edge rails, the type of rail still used today, albeit made from steel.

'Wylam Dilly' and 'Lady Mary' followed 'Puffing Billy'. These used similar principles but benefited from Hedley and Hackworth's experience in building 'Billy'.

'Puffing Billy' in action, from Pangborn's The World's Rail Way (above); the original 'Puffing Billy' at The Science Museum in London (below).

GEORGE AND ROBERT STEPHENSON

It is at this point that George Stephenson, perhaps the most famous name in the history of railways, enters the story. The Wylam wagonway ran past the cottage where Stephenson was born in 1871 and he must have seen Hedley's engines at work. Having successfully repaired a Newcomen pumping engine at Killingworth colliery, north of Newcastle, Stephenson was appointed enginewright there in 1812. He built his first locomotive, 'Blucher', for the colliery owners in 1814. The engine could pull a load of 30 tons up a gradient of 1 in 450 at 6 km (4 miles) per hour, but it still used a single-flue boiler, which had already been proved to be inferior to Trevithick's return-flue design.

Stephenson soon attracted the attention of colliery owners across North East England. He was engaged to build railways from mines to the ports, largely using stationary engines to haul coal wagons on rope-worked inclines. One such railway, for the Hetton Coal Company, opened in 1822. It used rope inclines on the hilly sections and five locomotives for the flatter sections of a 13 km (8 mile) railway to the port at Sunderland. This was the first railway to be worked from the start without animal power.

By far the most ambitious railway project to date was the 40 km (25 mile) Stockton and Darlington Railway. On 19 April 1821, the Stockton and Darlington Railway bill received royal assent, authorizing construction of the line. On the same day, Edward Pease, a Darlington businessman and major shareholder in the S&DR, met George Stephenson and the two men discussed the merits of the steam locomotive – all mention of which had been omitted from the Stockton and Darlington Railway Act. Stephenson was asked to make a fresh survey of the proposed route and in January 1822, was appointed engineer for the works. At a formal ceremony on 23 May, the first rails were laid, at a distance of 1422 mm (4 ft 8 in) apart, effectively setting the 'standard gauge' for most subsequent railways. Exactly one year later, an amendment to the Act received royal assent, authorizing some deviations from the original route of the line, but more significantly providing powers for the line to be worked wholly or in part by steam locomotives. However, it wasn't until 16 July 1824 that the Railway asked Robert Stephenson & Co to tender for the supply of two locomotives. The locomotive manufacturing company was set up in the name of George's son, Robert, although Edward Pease and George himself were also partners.

On 27 September 1825, 'Locomotion No. 1' headed a train consisting of 34 wagons, one coach and the locomotive's tender from Shildon to Darlington and Stockton, with George Stephenson at the controls. The railway age had begun.

'Locomotion No. 1', from Ahrons' The British Steam Locomotive, 1825–1925.

THE STOCKTON AND DARLINGTON RAILWAY

The first Stockton and Darlington Railway bill came before Parliament on 5 February 1819, but was opposed by landowners and the trustees of nearby turnpike roads. The bill failed at its second reading. Undeterred, the promoters worked to overcome the objections, refining the route where necessary. A new bill was presented in 1821 and received royal assent on 19 April 1821.

George Stephenson was appointed to make a fresh survey to ascertain that the route approved by Parliament was practical. Both the Act and the company's calculations had assumed that horsepower would be used. In January 1822, Stephenson was appointed engineer for the line and was directed to proceed with the unaltered parts of the route, while parliamentary approval was sought for any changes.

In November 1822, the 13 km (8 mile) Hetton Colliery Railway opened near Sunderland, using self-acting inclines, stationary engines and locomotives to move coal wagons. This gave the directors of the Stockton and Darlington an opportunity to see a machine-worked line as they prepared to put the changes to their railway before Parliament. Included in the S & D proposals was the use of locomotives, stationary engines and horses. The bill was enacted on 17 May 1823.

Robert Stephenson & Co were asked to provide stationary engines for the inclines at Etherley Hill and Brusselton Hill and on 16 July 1824, the company was asked to tender for the supply of two steam locomotives. The price quoted was £500 each.

The railway opened on 27 September 1825. The board and guests gathered at Brusselton to see coal wagons raised and lowered on the inclines before being attached to 'Locomotion' and a carriage. Guests clambered aboard empty wagons and the train set off, reaching a speed of 24 km (15 miles) per hour approaching Darlington. Six coal wagons were detached (the coals being distributed to the poor) and two wagons containing a brass band were coupled up. Three hours and seven minutes after leaving Darlington the train arrived at Stockton, the band played 'God Save The King' and a salute, fired from seven guns on the company's wharf, heralded the start of the railway age.

Two contemporary views of the opening of the Stockton and Darlington, the first railway designed from the start to use steam locomotives.

OPENING OF THE FIRST ENGLISH RAIL-WAY BETW

J. R. Brown

KTON AND DARLINGTON, SEPT. 27TH 1825.

THE LIVERPOOL AND MANCHESTER RAILWAY

The Liverpool and Manchester Railway Company was formed on 24 May 1823 by a group of merchants and businessmen. Their objective was to connect the two cities so that goods and materials could be moved quickly and cheaply. After a flawed first attempt at a survey by William James, George Stephenson was appointed engineer in 1824. His survey was little better, as his son Robert had gone to work in South America and George couldn't do the calculations required; the first railway bill was thrown out by Parliament. The directors then turned to the famous Scottish engineers, George and John Rennie, who appointed Charles Vignoles to carry out yet another survey. This was accepted and the Railway's second bill received royal assent in 1826.

When construction started, the Rennie brothers demanded highly inflated fees for the work, so George Stephenson was reappointed principal engineer. His first task was the crossing of Chat Moss, an immense peat bog. Stephenson's idea was to 'float' the railway across the bog on tree branches and hedge cuttings spread along the route, then covered with a thin layer of gravel on which the railway was laid. The work was completed on 1 January 1830, when 'Rocket' hauled an experimental passenger train over the floating railway.

As the works neared completion, the matter of motive power had yet to be decided. At George Stephenson's suggestion, the company held open trials at Rainhill, where Robert Stephenson & Company's 'Rocket' proved the winner.

Months of trial running were carried out before the railway opened amidst great ceremony on 15 September 1830, with eight Stephenson locomotives in charge of the trains. It was on opening day that William Huskisson, a local MP and a fervent supporter of the line, was run over by 'Rocket', his leg being almost severed beneath the wheel. He was conveyed by 'Northumbrian' to the vicarage at Eccles, but died a few hours later.

Public service began the next day.

William Huskisson MP, the world's first railway casualty (right).

Two contemporary views of the opening of the L&MR (below).

Mount Olive Cutting on the L&MR, painted by Henry Pyall after an engraving by Thomas Talbot Bury (opposite).

A FIRST-CLASS TRAIN ON THE LIVERPOOL AND MANCHESTER RAILWAY 1833.

A SECOND CLASS TRAIN ON THE LIVERPOOL AND MANCHESTER RAILWAY 1833.

THE RAINHILL TRIALS

The rapid growth in trade in South Lancashire gave rise to ideas for a railway between Liverpool and Manchester. George Stephenson was appointed engineer for the line, which he proposed should be worked by locomotives travelling at speeds of up to 19 km (12 miles) per hour. Some of the board members felt that steam locomotives were neither reliable nor sufficiently powerful. Stephenson prevailed upon the directors to decide the matter by staging open 'trials', laying down conditions that each entrant should be able to meet. Competitors' locomotives had to be able to travel over a trial length of 2.8 km (1¾ miles), backwards and forwards, to a total of 56 km (35 miles) – the distance from Liverpool to Manchester – at a speed of not less than 16 km (10 miles) per hour. The weight of the locomotives was not to exceed 4½ tons and they must be capable of drawing three times their own weight. A prize of £500 was offered to the winner. The 'trials' were held at Rainhill, halfway along the line, beginning on 6 October 1829.

Ten machines were entered, but only five actually began the tests. First up was Robert Stephenson & Co's 'Rocket'. Weighing just over 4 tons, it hauled a train of about 13 tons at a speed of over 16 km (10 miles) per hour. Next up was 'The Novelty', built by Messrs. Braithwaite & Erickson, of London. Weighing a little under 3 tons,

'The Novelty' was first tested without loaded wagons and achieved an impressive speed of over 48 km (30 miles) an hour, after which the engine promptly broke down and its laden trial was postponed to the following day. During the first day there was also a demonstration of a 'manumotive' carriage propelled by two men and carrying six passengers, but reportedly it moved 'with no great velocity'.

The next day, 'The Novelty' hauled a load of more than 11 tons at a speed of 32 km (20 miles) per hour, but then wet weather prevailed and the trials were suspended.

On 8 October, the judges moved the goalposts. Now, amongst other changes, the engines had to make twenty trips on the trial run, a distance of 113 km (70 miles), with a break after the first ten trips. Stephenson's 'Rocket' again went first, travelling the first 56 km (35 miles) at an average of more than 18 km (11 miles) per hour and the second 35 miles at more than 19 km (12 miles) per hour, with a best performance of 26 km (16 miles) per hour. Braithwaite & Erickson then asked for time to prepare their engine to meet the new rules, so the competition resumed on 10 October. 'The Novelty' attempted a 'rehearsal' trip, but a steam pipe failed. To entertain the spectators while repairs were undertaken, Stephenson

RACE OF LOCOMOTIVES AT RAINHILL, NEAR LIVERPO

brought out 'Rocket' again and in one run, the unladen locomotive clocked up a top speed of 48 km (30 miles) per hour, equalling 'The Novelty' in its demonstration run on the first day. 'The Novelty' then made a number of trips for the spectators, achieving a maximum speed of 35 km (22 miles) per hour when pulling a load of 10 tons.

On 13 October, day six of the trials, Timothy Hackworth's 'Sans Pareil' made its first appearance. Having been involved in the early steam locomotive experiments at Wylam, Hackworth had become resident engineer at the Stockton and Darlington Railway. He established the railway's repair and maintenance facilities at Shildon, in County Durham, designing and building new, powerful locomotives for the line. Unfortunately, 'Sans Pareil' exceeded the maximum weight for locomotives for the trial, but was allowed to proceed with its first 113 km (70 mile) trial. It soon proved to be a very powerful engine, completing 40 km (25 miles) in just two hours, before it too suffered a mechanical failure.

Having been repaired, Braithwaite & Erickson's 'The Novelty' resumed the competition the following day, but after just 10 km (6 miles), it failed again and was withdrawn. So 'Perseverance', entered by Timothy Burstall of Leith, began its trials, but achieved a top speed of no more than 8 km (5 miles) per hour. 'The Cycloped', entered by a Mr Brandreth of Liverpool, was a machine powered by a horse. Never regarded as a serious contender, it achieved a top speed of only 8 km (5 miles) per hour. 'Rocket' was the only engine to have completed a 113 km (70 mile) run and Stephenson was declared the winner.

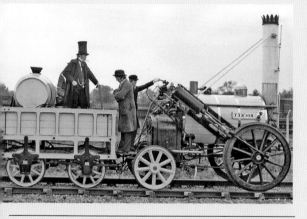

Replica of 'Rocket' at the Didcot Railway Centre (above).

A contemporary print of the Rainhill Trials (below).

ICH GEORGE STEVENSON'S ROCKET WON, 1829.

THE STEPHENSON'S SUCCESS AT RAINHILL

'Rocket' was a truly pioneering invention and it is worth looking carefully at what made it such a revolutionary design. In advising the board of the Liverpool and Manchester Railway to stage the trials, George Stephenson was likely to specify conditions that his own firm's products were able to meet. Nevertheless, 'Rocket' represented a huge leap forward in locomotive design.

Visually, 'Rocket' looked much closer to the familiar design of a steam locomotive. The cumbersome valve gear of earlier models had gone, replaced with cylinders mounted on either side, driving the wheels via coupling rods. The crucial improvement, however, was not visible as it was inside the boiler. The single flue used in previous locomotives had been replaced by an array

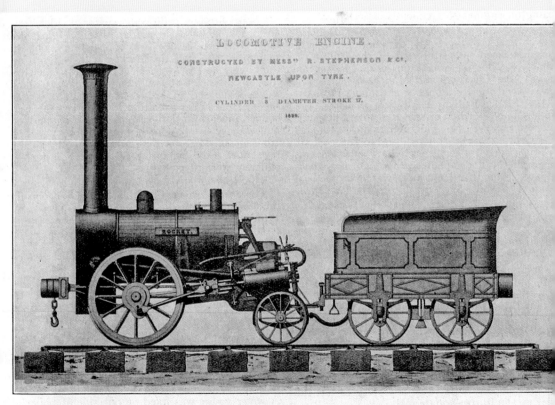

FIG. 6—THE " ROCKET," LIVERPOOL AND MANCHESTER RAILWAY, AS REBUILT AFTER 1829

This drawing is believed to be the one made to the order of the Directors of the Liverpool and Manchester Railway in 1836, when decided to sell that famous locomotive out of their service. It represents the engine after reconstruction, and shows that it essentially in its present form at that date. The drawing is particularly important in that it indicates the true form of the fire-

of twenty-five 75 mm (3 in) diameter copper tubes, through which the hot gases travelled, greatly increasing the heating surface from which to create steam. This wasn't Stephenson's invention: the concept came from Marc Seguin (1786–1875), a French engineer, while Stephenson was encouraged to use the design by Henry Booth, the treasurer of the Liverpool and Manchester Railway. The steam was drawn through the tubes by directing the exhaust steam from the cylinders up the chimney via a blastpipe (introduced by Trevithick around 1804). The harder the engine worked, the more it drew on the fire, thus generating extra heat in the boiler at exactly the time it was required.

The other important innovation was a water jacket around the firebox. This transferred extra heat to the boiler. It was perfected on later models when the firebox was completely enclosed within the boiler.

Robert Stephenson & Co built 'Rocket' in great secrecy at their locomotive works in Newcastle and tested it on the colliery railway at nearby Killingworth, before shipping it by sea to Liverpool.

George Stephenson portrayed at the height of his career in an engraving of 1830.

'Rocket' in her two states, clearly showing the improvements that led to her triumph in the Rainhill Trails (from a Scientific American article of October 1884).

THE ROCKET, 1829.

THE ROCKET, 1830.

This sketch of the Rocket I made at Liverpool on the 12th of September, 1830, the day before the opening of the Liverpool and Manchester Railway, while it remained stationary after some experimental trips in which George Stephenson acted as engine driver and his son Robert as stoker. JAMES NASMYTH.

IMPROVING THE DESIGN

Trevithick's Pen-y-darren locomotive of 1804 had two features that would become standard in locomotive design for the next 150 years. These were horizontal cylinders and a blastpipe, which directed steam up the chimney to draw the fire. These lessons, however, were quickly forgotten and early George Stephenson locomotives used upright cylinders and rocking levers, like the old-fashioned stationary beam engines.

To couple driving wheels together, first toothed gear wheels were tried and then a chain drive, until eventually the coupling rod was adopted as being the most efficient and by far the quietest method. Robert Stephenson's 'Rocket' used inclined cylinders (still not horizontal but clearly on the move) and a multi-tube boiler, which had 25 tubes through which the hot gases from the fire could heat the water.

The 'Northumbrian', the first locomotive with an integral firebox, in a drawing of 1831 (below).

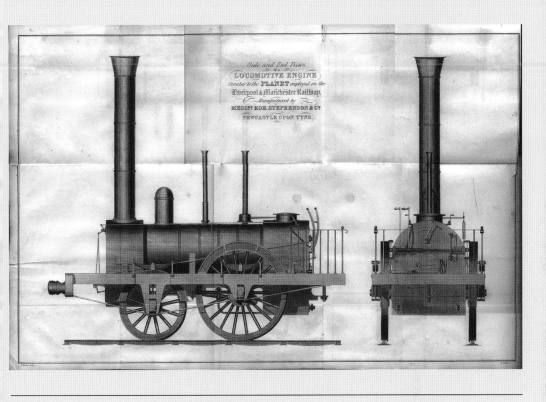

A Planet-style locomotive of the 1830s in an engraving by William Miller after J. Kindar (above).

The 'Northumbrian' that followed had two horizontal cylinders, albeit at the back of the locomotive driving the front wheels, a firebox enclosed within the boiler to deliver maximum heat from the fire and an opening smokebox at the front of the boiler. The smokebox housed the blastpipe and enabled the ash and cinders that accumulated in the fire tubes to be swept out at the end of each day's use.

In 1830, the cylinders inside the locomotive's frames were moved to the front of the engine in Stephenson's 'Planet' class locomotive. With all the features of the 'Northumbrian' incorporated as well, this was more or less the definitive design of the steam locomotive until it was superseded nearly 140 years later. It had taken just 26 years from Trevithick's first attempt at steam locomotion to arrive at this design.

One last great improvement was made to the steam locomotive. The superheater allowed steam to be heated above its boiling temperature, enabling more energy to be extracted from the steam in the cylinders and avoiding water condensing inside them. This came in the early twentieth century, when modern lubricants could withstand the higher temperatures.

BRUNEL AND THE ROUTE TO THE WEST

Isambard Kingdom Brunel had already established a reputation as a talented engineer when he was appointed chief engineer of the Great Western Railway in 1833. Probably his most controversial proposal was that the railway should be built to a gauge of 2134 mm (7 ft) – the 'broad gauge' – rather than Stephenson's standard gauge; but Brunel successfully petitioned the Board of the GWR, producing calculations to show that broad gauge trains would provide both a smoother ride and permit faster running speeds. Despite the rivalry between the two men, the first locomotive for the GWR came from Robert Stephenson & Co in Newcastle, the firm having already become established as the UK's leading locomotive builder. The 'North Star' was of a standard Stephenson design, originally built to a gauge of 1675 mm (5 ft 6 in) for the New Orleans Railway in America, but modified for the broad gauge and delivered to the GWR on 28 November 1837. 'North Star' worked the GWR's inaugural train from Paddington to Maidenhead on 31 May 1838.

Before the arrival of this first engine, Brunel appointed twenty-year-old Daniel Gooch to the post of Superintendent of Locomotive Engines. Gooch was born in Bedlington, Northumberland and had worked for Robert Stephenson & Co. While Brunel was undisputedly a great civil engineer, as a mechanical engineer, he was somewhat lacking. He laid down design rules for a batch of locomotives that he put out to tender with various manufacturers. It was Gooch's unenviable task to try and run a railway service using these locomotives and, as he said in his memoirs, 'I felt very uneasy about the working of these machines, feeling sure they would have enough to do to drive themselves along the road.'

In the GWR's early days, 'North Star' was the only reliable motive power which Gooch had at his disposal, the Brunel designs constantly breaking down and being unavailable for service. 'Morning Star' arrived from

Robert Stephenson in early 1839 and Gooch quickly ordered a further ten 'Stars', which were delivered between July 1839 and November 1841. These then provided the backbone of the service, the Brunel designs being quietly left to gather dust at the back of the shed. At the same time, Gooch's own design, which became known as the 2–2–2 'Firefly' class, was ordered

from seven different builders, eventually totalling 62 locomotives and cementing his status as a first-class locomotive engineer.

Gooch established a locomotive repair shed at Swindon, Gloucestershire, which opened in 1843 and employed 200 men. By 1846, Swindon had begun to manufacture the GWR's own broad gauge locomotives, the first of which was the 'Great Western'. By 1851, the locomotive works employed more than two thousand men and was producing about one locomotive every week.

Britain's railway network grew from just 40 km (25 miles) in 1825 to 9,800 km (6,100 miles) in 1851. Manufacturers all over the country were building locomotives large and small.

A replica of 'North Star', built in 1925 at the GWR Swindon Works and now on display at Swindon Steam (left).

The locomotive shop, GWR Swindon Works, 1881, photographed by R.H. Bleasdale (right).

The working drawings for the rebuilt 'North Star', reproduced in E.L. Ahron's classic The British Steam Locomotive, 1825–1925 of 1927 (below).

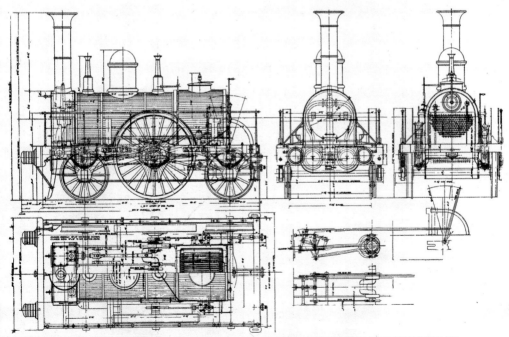

FIG. 43—GENERAL ARRANGEMENT OF THE "NORTH STAR" G.W.R. OF 1837, AS RECONSTRUCTED AT SWINDON WORKS IN 1925
Reproduced from drawing by courtesy of Mr. C. B. Collett, Chief Mechanical Engineer
For a detailed description see THE LOCOMOTIVE, Vol. XXXII, 1926, page 10 *et seq.*

BRUNEL AND THE GREAT WESTERN RAILWAY

The merchants of Bristol were anxious to cement their position as England's major port for the Americas and were aware that Liverpool was becoming more attractive to shipping. Their ambitious proposal was to construct a railway to London to rival any in the UK at the time.

The Great Western Railway was granted its Act of Parliament in 1835, having already appointed the talented Isambard Kingdom Brunel (1806–59) to be its engineer in 1833, when he was just 29. Brunel led the company to make two decisions – firstly, to use his broad gauge of 2134 mm (7 ft) as opposed to the 'narrow' gauge of 1435 mm (4 ft 8½ in), and secondly, to use a route to the north of the Marlborough Downs for the main line. While this line of route passed no major towns, it offered the possibility of connecting branches to Oxford and Gloucester. Brunel particularly preferred it because of 'the superiority of its levels and the ultimate economy of working steam power upon it'.

The first part of the line opened between London and Maidenhead on 4 June 1838 and reached Twyford the following year. Bristol to Bath opened on 31 August 1840 and the following December, the line eastwards reached the outskirts of Swindon. The through route to Bristol was completed with the construction of the 3 km (2 mile) Box Tunnel in 1841, which was the most difficult feat of engineering on the whole line, taking five years to build and costing the lives of around a hundred navvies. It is said that each year – at dawn on 9 April – Brunel's birthday, the sun shines from one end of the tunnel to the other. Another of Brunel's masterpieces on the GWR was his 'flat arch' bridge over the Thames at Maidenhead. This has two spans, each of which is 39 m (128 ft) in length and rises just 7 m (24 ft) at the centre.

It is not surprising that Brunel frequently said of the Great Western Railway during its construction: 'this is the best work in England'.

Isambard Kingdom Brunel in the famous 1857 photograph by Robert Howlett (left).

'Rain, Steam and Speed', J.M.W. Turner's painting of a GWR express crossing Brunel's Maidenhead Bridge in 1844 (below).

No. 1 Tunnel, from an 1846
book of engravings of the
Great Western Railway by
John Cook Bourne (left).

THE BATTLE OF THE GAUGES

George Stephenson's early locomotives were built for wagonways of 1422 mm (4 ft 8 in) gauge. When Stephenson joined the Stockton and Darlington Railway in 1822, he recommended the gauge with which he was familiar. This gauge later had 13 mm (half an inch) added to it to allow greater clearance round bends.

The chief engineer of the Great Western Railway, Isambard Kingdom Brunel, surprised the GWR board by recommending a broad gauge of 2134 mm (7 ft), which was later similarly stretched by 6 mm to 2140 mm (7 ft ¼ in) for clearance reasons. Brunel calculated that a broad gauge would permit higher speeds and provide a smoother ride and that the reduced friction should produce fuel economies. The locomotives would have a lower centre of gravity for added stability because the boiler could be slung between the frames, rather

than on top of them. Detractors said that the broader track needed more land and wider tunnels, viaducts, embankments and cuttings, thereby increasing capital costs and where the gauges met, passengers would need to change trains.

A notable example of this problem arose when the GWR built the line from Bristol to Gloucester in broad gauge in 1844. Passengers had to change at Gloucester, while their luggage and all goods and parcels had to be transferred by porters from train to train in specially built trans-shipment sheds.

The Government set up a Gauge Commission in 1845 to decide once and for all between the two factions. A trial was organized between a broad gauge train headed by 'Firefly' class 2–2–2 'Ixion' and a Stephenson long-boilered 4–2–0

on the standard gauge. The broad gauge locomotive was superior in all the tests, but this counted for nothing. At that time, 87.4 percent of track was laid to the standard gauge and the Commission came out in its favour. This became law on 18 August 1846. Gradually the GWR added an extra rail to its routes permitting through running, but it wasn't until May 1892 that the Great Western was fully converted to standard gauge and Brunel's broad gauge was consigned to history.

A broad gauge 'Parliamentary' train at Bristol around 1847 (left).

'Emperor', a GWR Dutchman broad gauge 4–2–2 locomotive built in 1880 (right).

The last broad gauge train to the West – the GWR 'Cornishman' photographed on 20 May 1892 (below).

' 'The Zulu' – Summer 1891', painted by Barry Freeman. During the final year of broad gauge era on the Great Western, 'Inkermann', one of Sir Daniel Gooch's famous 'Rover' class 4–2–2 locomotives heads the 'up' 'Zulu' past Worle Junction, Somerset. The class was never numbered, and the locomotives were known only by their names. All were scrapped following the abolition of the broad gauge in May 1892.

THE FORMATIVE YEARS,
1851–1895

The second half of the nineteenth century saw a number of crucial advances in locomotive design, leading to impressive improvements in the speed, strength and safety of steam locomotives. At the same time, the bewildering mix of railway companies running trains across the country began to merge with each other, leading to a less chaotic pattern of railway line construction and operation and the emergence of a recognizably modern national railway network. Yet this was also an era that witnessed the first of the great railway 'races', as the companies operating the eastern and western routes, between London and Scotland, competed to offer ever faster services.

THE FORMATIVE YEARS, 1851–1895
THE FASTEST TRAINS IN THE WORLD

In 1851, one British railway company was running trains at average speeds that none of the others could approach, let alone match; this company was the Great Western. Thanks to the superbly designed locomotives, the broad gauge track of 2140 mm (7 ft ¼ in) upon which they ran and the very lightly graded and gently curved line which joined London's Paddington station to Bristol, the 4–2–2 'Iron Dukes' were the racers of their day. A run from Paddington to Didcot, over 85 km (53 miles), was made at an average speed of 108 km (67 miles) per hour in 1848 and this remained a world record for more than 30 years.

turned over onto its side – unlike those locomotives involved in accidents on standard gauge railways, where this was almost always the outcome, often fatally injuring the crew. The Great Western Railway, at this time and for many years afterwards, was not only the fastest, it was also the safest railway in operation.

Sir Daniel Gooch in an 1883 photograph.

The man responsible for choosing the width between the rails and for laying out the route was Isambard Kingdom Brunel, one of the most famous names in engineering history. Because this piece of railway civil engineering is so level it is often referred to as 'Brunel's billiard table'. Yet Brunel had a weakness: his locomotive designs were poor, so poor in fact that they could barely haul themselves along, let alone tow a rake of carriages or freight wagons. Another personality was required to make the railway a success.

'Tornado', a Gooch GWR 8-footer broad gauge locomotive built in 1849.

Daniel Gooch was only twenty years old when Brunel appointed him Superintendent of Locomotive Engines to the Great Western Railway and it was Gooch's locomotive designs which provided the steam generation and power necessary to run at sustained high speeds. Because the broad gauge enabled all the mechanical parts of the locomotive to be mounted between the wheels, the engine's centre of gravity was very low; so if a broad gauge engine happened to leave the rails at speed (which was a rare occurrence) it almost never

Gooch on the footplate of the broad gauge 4–2–2 GWR express passenger Iron Duke class locomotive 'Lightning', shortly after it was built at the Swindon works in 1847 (right).

The GWR coat of arms (far right).

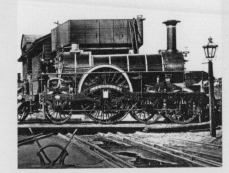

TRYING TO CATCH UP

All the other railway companies in Britain used the standard gauge of 1435 mm (4 ft 8½ in) between the rails. The narrowness of the track made their locomotives liable to topple over when negotiating bends, so engineers tried various methods to lower them, none of which was entirely successful. One approach was to mount the large diameter driving wheels at the back, behind the firebox, which enabled the boiler to be slung beneath the level of the axle, with smaller wheels supporting the weight of the locomotive. Unfortunately, these engines, designed by Thomas Crampton, tended to damage the track and they soon disappeared into obscurity.

When it was realized that standard gauge locomotives could never be as low-slung as their broad gauge rivals, improvements in design were focused on making the engines both more powerful and more efficient. This was accomplished in a number of ways.

THE TRANSITION TO COAL

The type of fuel burned in British steam locomotive fireboxes had always been coke, which is coal with all the volatile liquids and gases distilled from it. There was a ready and plentiful supply of coke, because it was a left-over by-product of the gas industry – coal gas being used to light houses and streets in towns throughout the country.

But coke was more expensive than coal and also a lot of the available thermal energy had already been removed, so a locomotive, which could burn, coal would be both cheaper to run and would generate more steam for the same volume of fuel. The problem was that the gases boiled from the coal shot straight through the boiler tubes and up the chimney without being properly burned. The solution was the fitting of a 'fire brick arch' in the firebox, which forced the gases to negotiate a longer journey before they entered the boiler tubes, giving them time to burn completely and release all their latent energy before being exhausted into the air. A deflector plate was also added to the firedoor, forcing air entering the furnace this way to follow the same elongated path.

Though broad gauge was considered safer than standard gauge, accidents still happened. The locomotive pictured here, 4–4–0 'Rob Roy', was pulling the up Mail from New Milford when it ran into the back of a cattle train on 5 November 1868. The broad gauge line was operated under the time interval system and the cattle train had stopped in the section. Thirty-six cattle and 8 men travelling in the rear vehicle were killed.

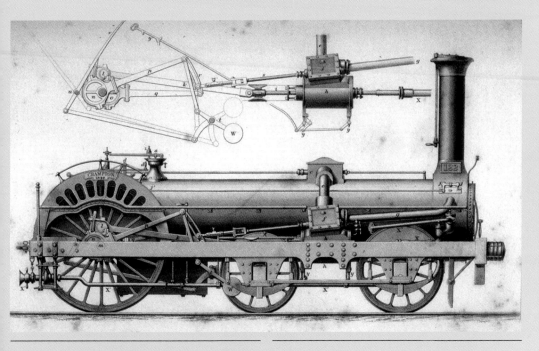

A Crampton locomotive in an 1848 engraving. Crampton's designs had a low boiler, large driving wheels and a single driving axle placed behind the firebox, all of which helped to give this design a low centre of gravity, making it safer at high speed even on standard gauge tracks.

The positioning of the deflector plate and brick arch (below).

This innovation was made right at the beginning of the period which we are discussing and tests were conducted by Matthew Kirtley, Mechanical Engineer for the Midland Railway for several years. By 1859, the design was perfected and all locomotives from this date burned coal instead of coke. This introduced new disadvantages, however, notably the clouds of black smoke which were regularly emitted from the chimney and the build-up of soot in the fire-tubes and smokebox, both of which required regular cleaning, a chore for the engine maintenance workers for the next hundred years.

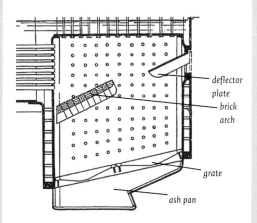

deflector plate

brick arch

grate

ash pan

EXPRESS LOCOMOTIVES

Throughout the period 1851 to 1895, express locomotives gradually became larger and more powerful. The fastest trains had always been hauled by engines, which were so-called 'single drivers', meaning that they had just one driven axle, with two large wheels mounted on it. The large diameter gave a high gearing, which meant that the engine could travel very fast. Some engineers were reluctant to use coupled driving wheels on high-speed trains, as they thought that the flailing coupling rods would present a problem. Single drivers were in use right up to the end of this period, the prime example being the 'Stirling G' class 2.4 m (8 foot) single drivers used on the East Coast Main Line from London King's Cross to York.

On the London and North Western Railway, Francis Webb designed the 'Precedent' class 2–4–0 for use on the West Coast Main Line from London Euston to Crewe and Carlisle. These had four driven wheels and turned out to be just as adept at running express passenger services as the single drivers, even surmounting the formidable Shap summit between Preston and Carlisle in some style. Later, when Webb introduced his 'compounds', these proved such a flop that the Precedents were often brought back into service.

The London, Brighton and South Coast Railway used William Stroudley's 'B1' class 0–4–2 locomotives on their express services. These had the four-coupled driving wheels at the front, leading the engine and many thought that they would not hold the rails, but they turned out to be very fast and efficient, hauling expresses from London to Brighton for more than twenty

years. Gladstone is the most famous example of this type and it is preserved today at the National Railway Museum at York.

The Great Western Railway was still using the broad gauge at this time and the express locomotives designed by Daniel Gooch continued to run from Paddington to Bristol and beyond for forty years or more. The 'Iron Duke' class 4–2–2 locomotives had been around since 1847, but were not withdrawn until the GWR was forced to abandon the broad gauge in 1892.

Glasgow and South Western Railway 4–4–0 express locomotive No. 74, October 1901.

Great North of Scotland Railway 4–4–0 express locomotive No. 115, August 1900.

Great Northern Railway 4–4–0 Irish Express locomotive No. 131, 'Uranus', November 1901.

North Eastern Region 4–6–0 express locomotive No. 2111, 1906.

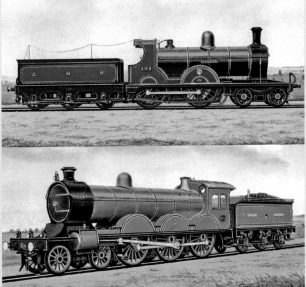

Great Northern Railway, Stirling 8-foot 'No. 1', leaves Kings Cross with a Scotland-bound express (left).

BRAKING SYSTEMS AND SAFETY

As could be expected, there were a number of accidents on the railways in the second half of the nineteenth century. When speeds were low and the trains lightly loaded, brakes were not considered to be of too much importance. The engines of this period typically had no brakes at all, there being a hand-operated screw brake on the tender, which could be wound on when the situation demanded.

Freight trains were unbraked apart from a ballasted brake-van at the rear in which the guard could apply his brake when the driver demanded, using a signal from the engine's whistle.

Continuous brakes fitted to passenger trains operated throughout the nineteenth century, but they were not of the fail-safe, or so-called 'automatic' variety, in which the brakes are automatically applied when there is a failure. Instead, they only operated when air, or a vacuum, was pumped into or from the train pipe linking all the carriages. If the pipe was broken, during an accident for example, then the brakes were released and the carriages were free to continue on their way.

Accident inspectors throughout the period constantly urged the fitting of continuous automatic brakes to all passenger trains, but many companies refused, citing the financial cost as being too high and the number of accidents to be extremely low.

Two automatic systems were available: the vacuum brake held the train pipe connecting the carriages at a very low pressure, which kept the brakes off. When the pipe was severed, air was admitted and the brakes went on. The Westinghouse system was a high-pressure air brake: it used air pressure in the train pipe to hold the brakes off, and again, if the pipe became broken or detached, the pressure was lost and the brakes were applied.

The final straw for the inspectors was the Armagh disaster of 1889, when the engine of a train carrying schoolchildren stalled near the top of a steep hill. The train was divided so that the engine could take the first five carriages to a siding, then return for the remainder. Unfortunately, the locomotive rolled backwards on starting, nudging the rear portion of the train sufficiently for it to overcome the guard's brake and it ran away down the hill, smashing itself to pieces against the engine of a following train, killing eighty, most of them children. Within the year, the law was passed which made the automatic brake compulsory on all passenger trains in Great Britain.

At the Newark brake trials of June 1875, the train entered by the London and North Western Railway broke in two when the Clarke & Webb chain brake was applied suddenly, showing the danger of trains separating under sharp braking.

A Pullman coach on the Midland Railway in 1874. Before the introduction of continuous brakes, these carriages were individually braked by hand.

On 12 June 1889, 80 people died and hundreds were injured in Ireland's worst railway disaster when the brakes were unintentionally released on a string of ten stationary carriages. Many of the dead were children on a Methodist Sunday School excursion and because it was common practice at the time to lock carriage doors on passenger trains, they had little chance of escaping before the carriages plunged down the embankment.

An illustration from the January 1903 issue of Railway Magazine, showing the vacuum brake system, required by law to be installed on all British trains after the Armagh diaster of 1889.

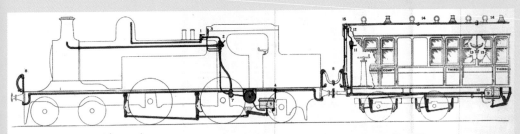

DIAGRAM ILLUSTRATING GENERAL ARRANGEMENT OF

VACUUM AUTOMATIC BRAKE

ON ENGINE, CARRIAGE AND VAN.

TABLE OF REFERENCE.

1—Stop Valve.	6—Vacuum Chamber.	11—Van Valve.
2—Combination Ejector.	7—Drip Trap and Valve.	12—Gauge.
3—Duplex Gauge.	8—End Pipes and Coupling.	13—Passenger Communication Handles.
4—Syphon.	9—Cylinder.	14—Roof End.
5—Cylinder.	10—Syphon.	15—Air Valve.

GETTING THE MOST FROM THE FIRE

With the use of coal, bigger and more efficient furnaces were required in order to get as much heat as possible from the burning fuel and into the boiler. In 1860, a Belgian called Alfred Belpaire invented a new type of firebox for locomotives. It had a large rectangular shape, which was more difficult to join to the boiler than the existing rounded designs, but had great advantages in the larger surface area it presented to pass heat from the furnace and into the water that surrounded it. It was so efficient that many engineers used it, despite its complexity, and its distinctive shape can be spotted on several of the later Victorian locomotives and even more from the twentith century.

INCREASING BOILER PRESSURE

In the middle part of the nineteenth century, boiler pressures were around the 120 pounds per square inch (psi) mark. If this figure could be increased, not only would the steam engine be more powerful, because a bigger force would be exerted on the pistons, but because of the laws of thermodynamics, the whole steam cycle would be more efficient and less coal would be consumed for the same power output. Engineers therefore knew that greater boiler pressure was desirable, but it could only be achieved if the boilers themselves could be made stronger. The weakest part of a locomotive boiler is around the firebox, where there are flat surfaces exposed to the full pressure of the boiler trying to blow them apart. These forces are repelled by hundreds of firebox 'stays' – steel bolts riveted at each end all around the sides and top of the firebox. After a while, they become thin and weak, as they are eroded by the scalding hot water and steam and they then all have to be replaced – no easy job, but the alternative is that the firebox collapses violently, at risk of killing or maiming the footplate crew.

As technology and materials improved rapidly throughout the period, boiler pressures gradually increased. The Midland Railway's 156 class 2–4–0 locomotive of

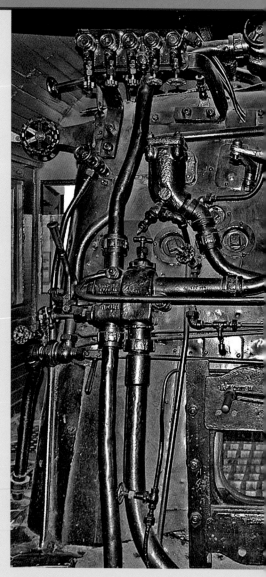

A Belpaire firebox on PRR Consolidation No. 2846 in the Railroad Museum of Pennsylvania, showing the square shape and intricate pipework (above).

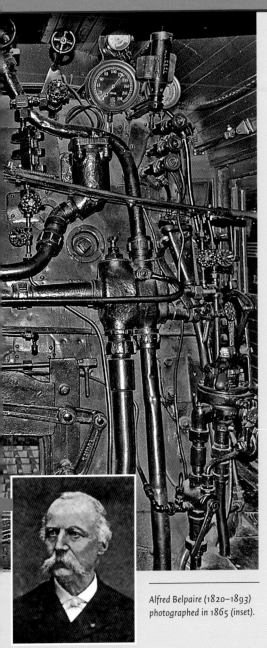

Alfred Belpaire (1820–1893)
photographed in 1865 (inset).

1866 boasted 140 psi in its boiler, as did Patrick Stirling's G class 4–2–2 of 1870, which hauled passenger expresses from London to Edinburgh for the Great Northern Railway. These engines were produced over a 25-year time span and the last ones had boilers rated at 160 psi. In the 1870s, Francis Webb of the London and North Western Railway started fabricating boilers using steel rather than wrought iron, claiming that they lasted longer and could reliably confine steam at higher pressures. By 1895, 175 psi boilers were the norm and pressures continued to increase right up to the middle part of the twentieth century, when 250 psi gave large locomotives like the 9F 2–10–0 their immense power.

GETTING THE MOST FROM THE STEAM

In a steam engine, the way that the steam is controlled, as it flows to the cylinders and then is exhausted through the chimney, is crucial to its performance and economy. The device which performs this task is called the 'valve gear' and its evolution from rudimentary form to the complex devices perfected during the 1850s enabled the locomotive to cope with all the varying conditions of load, gradient and speed, which it would encounter during its working life.

The earliest valves simply applied the high-pressure steam to the piston throughout the length of its stroke, the driver determining the rate at which the steam was admitted using the regulator. While this gave full power when required, it was extremely wasteful, as a whole cylinder-full of fresh steam was lost to the atmosphere every time the exhaust valve opened. The first innovation was therefore the ability to apply a 'cut-off' to the steam, so that it was only fed to the cylinder for approximately 20 percent of its travel. After this point, the inlet valve closed and the steam would expand against the piston, driving it along with reduced force, but using far more of the energy it contained and only drawing one-fifth as much from the boiler. The ability to change the cut-off while the engine was in motion enabled the driver to let in the steam for just the right amount of time according to the work required to surmount a hill, or haul a heavy load, or cruise along a level piece of track.

The other breakthrough was to be able to change from full forward gear, through increasing stages of cut-off, to the point where no steam was used and then to continue through to full reverse gear, all using one control and in a continuous sequence. This enabled the steam to be used as a brake, the pistons acting against the motion of the wheels and so slowing the locomotive down in an emergency. When danger was spotted, the fireman screwed down the hand brake on the tender, while the driver frantically wound on the reversing handle, the wheels squealing as they bit into the track – the famous sound beloved of movie directors shooting runaway trains.

The first type of valve gear to incorporate these features was invented at the works of Robert Stephenson & Co, the famous locomotive engineers and it was called 'Stephenson's Link Motion'. This became the most widely-used type in the second half of the nineteenth century. Its only drawbacks were that a considerable effort was required by the driver to change from forward to reverse and it took up a good deal of room between the engine's frames.

A Belgian called Egide Walschaerts developed a very similar valve gear to the Stephenson, which could be mounted entirely on the outside of the frames. This took a long time to gain popularity, but became the valve gear of choice for most steam engines built in the twentieth century.

Egide Walschaerts.

The valve gear of a GWR 4–6–0 6000 'King' class express locomotive.

The Walschaerts gear mechanism of the LMS 'Princess Coronation' class 6233 'Duchess of Sutherland'.

LMS Stanier 'class 5' 4–6–0 No. 44767 fitted out with a Stephenson valve mechanism as a trial to improve the original design of the 'Black Five.

COMPOUND LOCOMOTIVES

Up until the late 1870s, all steam locomotives were of the 'simple expansion' type, in which the steam is used just once and then is exhausted into the atmosphere, giving the characteristic 'chuff' sound. Engineers had long known that if you use the steam twice, you can recover more energy from it and for years many steam ships carried what were called 'compound' engines. In these, the steam first enters one, or more, small high-pressure cylinders, then is fed into larger low-pressure cylinders, gradually increasing in size, until all the available energy is used up. However, installing such an engine in the tight space of a locomotive was not so straightforward.

The first person to use compounding in Britain was Francis William Webb (1836–1906), the autocratic Chief Mechanical Engineer of the London and North Western Railway, who introduced his 'Experiment' class in 1882. However, his design of compound engine was quite bizarre and not entirely successful. It utilized two high-pressure cylinders, one on each side in the usual place for a locomotive. The exhaust steam from these was collected in a large chamber, then fed to a large single low-pressure cylinder mounted between the frames. The high-pressure and low-pressure cylinders drove separate axles, which were not linked together using connecting rods and there are many stories of Webb compounds floundering helplessly while their two sets of driving wheels turned in opposite directions!

The most successful application of compounding on British railways was by the Midland Railway which, in 1902, introduced the '4P' class, designed by Samuel Johnson, with a single, high-pressure cylinder between

N&NWR 8-coupled 4-cylinder compound coal engine No. 1881,
illustrated in the June 1903 issue of Railway Magazine (top).

GWR 4–4–2 express locomotive No. 103 of a De Glehn compound design,
Railway Magazine, November 1905 (above).

the frames and two, low-pressure cylinders on the outside – the very opposite to Webb's design. Two hundred and forty of these were built and gave excellent service, the final example not being withdrawn until 1961.

So, why did other companies not use this type of engine? Although they were more efficient and gave better fuel economy than the simple type, they were complicated to drive, especially when starting and the skill of the driver was a large ingredient of how well they performed.

THE SEARCH FOR MORE STABILITY

If the confined width of standard gauge track prevented engineers from lowering the centre of gravity of their locomotives, another way had to be found to make them more stable, especially in the bends. The four-wheeled bogie, pivoted at the rear of its carrying truck and used to guide the locomotive into corners, was invented in Britain and patented by Chapman in 1812, but it was in America where its use became almost universal, so much so that the 4–4–0 wheel arrangement is often referred to as the 'American'. In Britain, it was ignored until the early 1860s, when its advantages for smooth and high-speed running on less than perfect track became apparent. The 4–4–0 became the standard wheel arrangement for express locomotives in the later part of the nineteenth century and the early twentieth.

Preserved Midland Railway No. 1000 in steam at the Rainhill Trials 150th anniversary in 1980 (above)

The Walschaerts valve gear on SR 4–4–0 'V' class No. 925 'Cheltenham', National Railway Museum, York (opposite).

GETTING WATER INTO THE BOILER

The locomotive boiler has constantly to be supplied with fresh water to replace the water which is turned into steam and used to power the wheels. Up until 1860, the only way this could be done was using a mechanical pump driven off one of the axles, or off the piston rod. It was therefore impossible to replenish the boiler while the engine was stationary. French balloonist, Henri Giffard, invented a device which used the power of the steam itself to force the water into the boiler – this was called the 'live steam injector' and it was fitted to nearly all locomotives from this time, enabling them to keep the boiler topped up and thus operating safely. Another improvement was the 'exhaust steam injector', which used power from the steam after it had been through the cylinders – that would otherwise have gone to waste.

Henri Giffard, and his live steam injector.

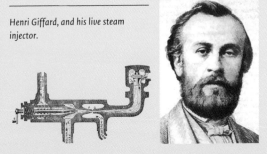

THE RACES TO THE NORTH

Although the early railways were clustered around the industrial north of the country, the success of the Liverpool and Manchester made people realize that the capital, London, should be connected by railway with the rest of the country as soon as possible. There were two main routes out of London to the north in the early years.

The first was from Euston to Birmingham on the London and Birmingham Railway, which had been engineered by Robert Stephenson. North from Birmingham, the Grand Junction line met the old Liverpool and Manchester Railway at a point between those two cities. These companies amalgamated into the London and North Western Railway (LNWR), which extended all the way to Carlisle. Beyond Carlisle, the journey into Scotland was continued by the Caledonian Railway (CR).

The second main route north was from King's Cross via the Great Northern Railway (GNR) to York, then on to Berwick-upon-Tweed via the North Eastern Railway (NER), then into Scotland and beyond with the North British Railway (NBR).

In the 1880s and 90s, these two routes vied with each other to provide the fastest services between London and Scotland. The first of these unofficial races started in November 1887, when the east coast railways teamed up to announce that they would run from King's Cross to Edinburgh in nine hours exactly. The LNWR, in partnership with the Caledonian on the west coast route, had always scheduled its 10 a.m. departure from Euston to arrive in Edinburgh at 8 p.m., an overall time of ten hours. The LNWR immediately responded by lowering their own timings to match those of the east.

LNWR No. 790 'Hardwicke' at the York Railway Museum (below).

Kinnaber signal box shortly before closure and demolition in October 1981 (inset).

They continued in this fashion, step by step, until the timings reached 7 hours 30 minutes for the west coast route and 7 hours 32 minutes via the east coast. Some of the travelling public were becoming nervous, mindful of the speeds which were required to perform these feats and the inherent dangers. The two sides called a truce and agreed on timings: 8 hours for the west coast and 7 hours 45 minutes for the east. This amicable state of affairs continued for the next seven years.

In 1895, a new series of races began, this time along an even greater distance between London and Aberdeen. In 1890, the Forth Bridge was opened across the Firth of Forth, north of Edinburgh. This gave the east coast partners a direct line to Aberdeen from King's Cross, which was 27 km (16½ miles) less than the 869 km (540 miles) from Euston. When, in the July of 1895, the west coast consortium rescheduled their 8 p.m. sleeper train to take 11 hours 40 minutes, this was five minutes less than the east coast's schedule. The east coast responded and a tit-for-tat battle ensued, both sides first cutting the timings by fifteen minutes, then a further twenty-five minutes and then by another fifteen.

The crux of the contest was reached, not at Aberdeen itself, but at Kinnaber Junction, some 61 km (38 miles) south, where both trains had to run over the same line for the last part of the journey. Whoever reached Kinnaber Junction first was declared the winner. The signalman at Kinnaber worked for the Caledonian Railway, which was allied with the west coast route and the west coast train won every night. But one morning in August 1895, on the penultimate night of the races, both trains arrived together and the signalman, inexplicably, allowed the east coast train to proceed first.

On the two nights of 21 and 22 August, the east coast reached Aberdeen in 520 minutes, including stops, a remarkable average of 97 km (60 miles) per hour. The west coast then completed their longer journey of 540 miles in 512 minutes, a phenomenal achievement.

But, once again, the races were becoming risky. Added to this was the inconvenience of arriving on a deserted station platform in the early hours of the morning – where was the advantage in that? Once again, agreement was reached over timings – the east coast would do the journey in 10 hours 25 minutes and the west coast over their longer route would take five minutes more.

The locomotives which took part in the races.

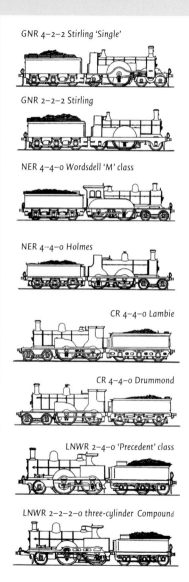

GNR 4–2–2 Stirling 'Single'

GNR 2–2–2 Stirling

NER 4–4–0 Wordsdell 'M' class

NER 4–4–0 Holmes

CR 4–4–0 Lambie

CR 4–4–0 Drummond

LNWR 2–4–0 'Precedent' class

LNWR 2–2–2–0 three-cylinder Compound

THE TURN OF THE CENTURY, 1896–1905

As passenger numbers grew and the railways settled into their role as the industrial and social backbone of Britain, now the richest and most powerful nation on earth, a new generation of steam locomotives appeared, fit to haul longer, heavier passenger and goods trains. The most successful of these locomotives would go on to serve the railways for more than half a century. This period also saw one last hurrah for the great Victorian railway visionaries, with ambitious, unrealized schemes, including Sir Edward Watkin's proposal for a channel tunnel, through which European trains would run onto the various railway lines of which he was a director, from the south coast of England to the industrial north.

Though this painting, by Malcolm Root, shows a scene from the 1980s, the locomotive is one of the great GWR 'King' class engines, preserved No. 6023 'King Edward II', built in 1930. The named train, 'The Cornish Riviera Express', however, started running in 1904, one of Britain's best-loved services as it took Londoners on holiday to Devon and Cornwall. The name came from a public competition in the August 1904 edition of Railway Magazine, the prize being three guineas. Among the 1,286 entries were two winning suggestions, 'The Cornish Riviera Limited' and 'The Riviera Express', combined as 'The Cornish Riviera Express', although railwaymen tended to call it simply 'The Limited'.

THE TURN OF THE CENTURY, 1896–1905
LOCOMOTIVES GET BIGGER

In 1895, express passenger locomotives were of varying types and wheel arrangements, but they shared one thing in common: they all weighed between 30 and 40 tons (not including the tender). On the Great Northern Railway, Patrick Stirling's 'G' class single-driver 4–2–2 was, by now, a very old-fashioned design. The Midland Railway and the London and North Western Railway both used 2–4–0 types for most of their passenger trains, the '156' class on the Midland and the 'Precedent' class on the LNWR. The London Brighton and South Coast Railway used the 0–4–2 B1 class 'Gladstone' type, while the Great Western Railway, on being forced to abandon the broad gauge in 1892, had 4–4–0 types developed by William Dean, which were quite advanced for their day.

Two typical GNR 4–4–2 express locomotives of the 1900s, Nos. 292 and 251.

But trains were getting longer and heavier; something bigger was needed for the job. The Midland Railway continued with small engines right up to the formation of the London Midland and Scottish Railway (LMS) in 1923. This policy was often criticized, but the railway made do by 'double heading' – using two locomotives at the front of nearly all its express passenger trains. The other companies strove for something larger, which could pull these greater loads on their own.

In 1898, the first locomotive with the 4–4–2 'Atlantic' wheel arrangement was introduced on the Great Northern. This had all the advantages of stability and cornering of the 4–4–0, but, with an extra single-axle truck (called a 'pony-truck') at the back the locomotive, could be made longer and heavier, enabling a large boiler to be fitted and providing extra adhesion from the greater weight on the driving wheels. In 1894, the first 4–6–0 had been introduced on the Highland Railway in Scotland, but this was for the hauling of goods trains only at that time – it gained the name 'Jones Goods', after the Engineer who designed it. It wasn't until 1900 that the 4–6–0 was first used on passenger trains, both on the Highland and on the North Eastern Railway.

In 1903, the first 2–8–0 was put in service on the Great Western Railway, one of the first acts by George Jackson Churchward (1857–1933) when he took up his position as Locomotive Superintendent for the company. This, too, was a freight design, the eight-coupled wheels giving great adhesion for the pulling of heavy trains. Churchward would go on to design the best 4–6–0 locomotives in the world for hauling express passenger trains, including the classic classes 'Saint' and 'Star'. The first 'Saints' were constructed in 1902–3 as experiments to test the design and the 'Stars' later evolved under Charles Collett (1871–1952) into the huge 'Castles' and 'Kings', which pounded the Great Western main line almost until the end of steam under British Railways in 1968.

The 4–6–2 ('Pacific' class) locomotive was also first seen on the Great Western Railway in 1908. It was called 'The Great Bear'. However, Churchward favoured 4–6–0 designs, as they placed more weight on the driving wheels, thereby delivering greater adhesion over the steep hills that were a feature of the line between Exeter and Plymouth.

So, 'The Great Bear' was soon rebuilt as a 4–6–0. The GWR never used a Pacific again, but the other large companies in Britain, even eventually the Midland (with its policy of using only 'small engines'), after it became part of the LMS, took to the Pacific layout in a big way and used it for their prestige express passenger services for many years to come.

GWR 4–6–2 Pacific
'The Great Bear',
No. 111, from a 1908
postcard (above).

GWR 4–6–0 'Hall'
class 'Olton Hall',
famously used to haul the
'Hogwarts Express' in the
Harry Potter films (right).

GREAT WESTERN CITIES

In September 1902, George Jackson Churchward, the Chief Mechanical Engineer of the Great Western Railway, installed the first version of his 'tapered' boiler on an existing 'Atbara' class 4–4–0 locomotive called 'Mauritius'. It was so successful that he ordered ten brand new locomotives to be built incorporating this superb steam-raiser and these became the 'City' class. A further ten of the old Atbara class were also rebuilt as Cities, making twenty in all.

The locomotives soon gained a reputation for speed. 'City of Bath' ran from London to Plymouth in July 1903 at an average speed of 63.2 mph, touching 81 mph at Chippenham.

But the most famous date in the history of locomotives, and in railway folklore in general, is 9 May 1904 when 'City of Truro' became the first steam engine in the world, indeed the first vehicle of any type, to travel at more than 160 km (100 miles) per hour.

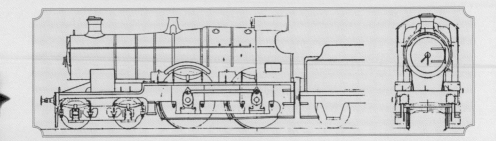

Front view and profile of a 'City' class locomotive (above)

'City of Truro' in full steam (left).

Nameplate of the restored 'City of Truro' (below right).

Plummeting down Wellington bank in Somerset, heading a train bound for Paddington, railway journalist, Charles Rous-Marten, who had been specially invited along on the trip, recorded a time of eight and four fifths seconds between two quarter-mile markers, giving a speed of 164.6 km (102.3 miles) per hour.

Although the boiler was ahead of its time, the construction of the rest of the locomotive was very old-fashioned, using a double-framed design, with the wheels sandwiched between the two frames and the connecting rods flailing around on their own on the outside. Almost all new locomotives at this time were being built with single frames, both the wheels and connecting rods being on the outside. This probably explains why only ten of this class were ever built, being superseded rapidly by the 4–6–0 'Saint' and 'Star' classes, which also adopted the tapered boiler pioneered so successfully in the 'Cities'.

THE MIDLAND RAILWAY'S 'SMALL ENGINE' POLICY

The Midland Railway, with the notable exception of the Settle and Carlisle line, had a network with relatively easy gradients and there was no great incentive to increase the size of their locomotives. The mainstays of the fleet were dozens of 2–4–0s and 4–4–0s designed by Matthew Kirtley, who was Locomotive Superintendent from 1844 to 1873. When he died, still in the job, hundreds of these types survived and remained in service right through the turn of the century, many hanging on until the formation of the London Midland and Scottish Railway in 1923.

His successor, Samuel Johnson (1831–1912), also had a long career with the Midland, not retiring until 1903. It could be argued that he was not exactly forward thinking. He spent most of his time designing goods locomotives, until in 1887 he reintroduced the single driver 4–2–2 layout of passenger engine, a type which was already decades out of date. The simplicity of the single driver had always been appealing, but its drawback was lack of grip with only two wheels driving. But the works manager at Derby, Francis Holt, invented the steam-powered sanding device. This blasted sand at high pressure between the driving wheels and the rails, controlled at an instant by the train driver. Ninety five single drivers were built over the next thirteen years, the most famous of which were the 115 class 'Spinners' with 2375 mm (7 ft 9½ in) driving wheels – single drivers may have been out of date, but no one could deny that they were beautiful.

However, Samuel Johnson was saving the best until last: right at the end of his tenure as Locomotive Superintendent, he released his masterpiece, the 'Midland Compound'. Johnson's '4P' class 4–4–0 introduced in 1902 was by far the most successful of the compound locomotives in Britain, ever. They were much more powerful beasts than the small engines, which were still struggling along, mostly double-headed, on the heavier trains of the early 1900s. A Johnson Compound could single-handedly draw a long express passenger train out of St Pancras and up the steep bank over the Regents Canal on its own. A total of 240 of these magnificent engines were built, and No. 1000 is preserved and can be seen in the National Railway Museum at York (see the picture on page 61).

Three examples of 4–4–0 engines from the 1910s: GWR No. 3310 'Waterford'; LBSCR No. 213 'Bessemer'; and GWR express locomotive No. 3473.

The Midland's small engine policy eventually caused a serious accident involving an express passenger train and loss of life. It wasn't on the level running grounds in the south of the Midland system but, as could have been predicted, up in the wild and storm-swept wastes of the north, on the Settle and Carlisle line, where the long, steep hills on either side of the route's summit would have been a challenge for even the most powerful engines of the day.

This mishap occurred in the early hours of Christmas Eve 1910, a few miles to the south of the summit, at the remote and windswept station of Hawes Junction. Directly as a result of the small engine policy, Hawes Junction was an extremely busy place – not for passengers (there were hardly any of those in the middle of the fells) – but for the pilot engines, which dragged the trains up from Carlisle in the north and were detached at Hawes where they were turned, then held, awaiting return to Carlisle and their next assignment. Meanwhile, the express trains which they helped up the gradient could continue on their journey unaided, freewheeling easily on the down slope heading southwards.

The signalman that night had no fewer than five pilot engines waiting to return to their respective depots – three for Leeds to the south and two for Carlisle. These two were turned so they faced north and then, coupled together, they drew onto the main line, waiting at the signal for their turn to proceed. There, according to the accident inspector, Major Pringle's report, they waited for around twenty minutes. They did not sound their whistles to remind the signalman of their presence; neither did they despatch one of the firemen to the signalbox to remind him in person, as required by the famous 'Rule 55': one of the fundamental safety rules on British railways for more than a hundred years.

The signalman had forgotten that they were there. When a northbound express following behind them was offered, he cleared his signals for it to pass. The two engines took this as their own signal and chuffed off into the night. The express, double-headed as was normal to get it up to the summit, thundered through Hawes a minute later at high speed. It caught the other engines up just as it emerged from Moorcock tunnel at 105 km (65 miles) per hour and crashed into the tender of the second engine at Grisedale crossing. The express turned on its left side against the embankment and coals from the engines ignited the gas cylinders in the coaches. The whole wreck was soon ablaze and twelve passengers died.

The signalman made no attempt to hide his mistake – the poor man had to live with it for the rest of his life. The inspecting officer laid most of the blame at his door, but he also severely reprimanded the two drivers of the light engines for not carrying out Rule 55. If, at any time during the 20 minutes they spent waiting at the red signal, they had done so, the crash would not have happened.

The accident at Hawes Junction on 24 December 1910 – the two engines of the express train with the smouldering wreckage of the carriages behind.

FAIRLIES, MALLETS AND GARRATTS

The Stephensons, father and son, arrived at the definitive design of the steam locomotive in 1830 with their 'Planet' class. From that date until the end of steam, 99 percent of the world's steam locomotives were to this layout, variations being minor ones such as the number of cylinders, their location inside or outside of the frames, the wheel configuration and, of course, the overall size. The three types discussed here formed some of the remaining one percent – engineers' attempts at improving the traditional way of doing things.

Scottish engineer Robert Fairlie (1831–85) patented his locomotive in 1864. He wanted a design which had all its wheels driven, so there were no carrying wheels to waste the tractive effort from the cylinders. His solution was a double-ended locomotive, with two boilers pointing in opposite directions, joined at the centre where the two cab roofs met. The two furnaces could, therefore, be fed by one fireman. His locomotives are mostly associated with the Ffestiniog Railway in Wales.

Swiss engineer, Anatole Mallet (1837–1919), patented his locomotive in 1884. He wanted to produce one which could carry a huge boiler, powering four cylinders, outside the frames, but which could still be used on lines with sharp curves. He did this by powering the front bogie (the two-axle truck fitted at the front of the locomotive to help the engine negotiate curves) of a 4–4–2 locomotive, so that eight wheels were driven and the powered bogie was

Robert Fairlie in a photograph taken in the late 1880s (left).

Ffestiniog Railway Fairlie double-ended locomotive 'Iarll Meirionnydd/Earl of Merioneth' (below).

hinged so that bends could be negotiated. In the USA, some of the largest and most powerful engines ever seen used the Mallet layout, such as the 4–8–8–4 Union Pacific, 'Big Boy'.

In the early 1900s, British engineer, H. W. Garratt (1864–1913), came up with an idea for a locomotive that had two powered bogies, with the boiler mounted between them and hinged at both ends. These could go round even sharper corners than the Mallet design and were ideal for use on tortuous routes, such as those found in Africa. They delivered immense power and could haul vast loads on narrow gauge tracks.

Mallet locomotives – a 1911 engine on a preserved railway in Alsace, France, and a Union Pacific 'Big Boy', on display at Pomona, California.

A heavy-duty Beyer-Garratt working in South Africa in the 1930s (below).

The original working drawings for the K1 Garratt locomotive (below).

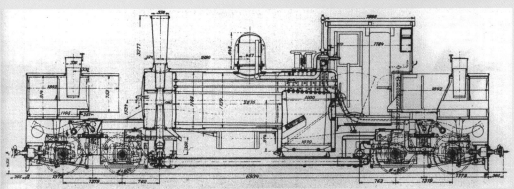

EDWARD WATKIN AND THE CHANNEL TUNNEL

Edward Watkin (1819–1901) was a railway visionary. He was also abrasive, vitriolic and enjoyed nothing better than to fight with anyone who opposed him. He worked for many railway companies during the latter half of the nineteenth century, eventually becoming a director of nine different ones, including the GWR, the Great Eastern and, most famously, the Metropolitan, where his rivalry with James Staats Forbes (1823–1904), who was chairman of the District line, lasted for more than thirty years. Much to the frustration of the authorities, and Londoners in general, their spat held back the linking of the two lines to form the Circle line for many years. When it was finally completed, a childish arrangement was put in place whereby trains of the Metropolitan company would run clockwise on the outer track, while those of the District would proceed anti-clockwise on the inner.

Edward Watkin in 1897

It was Watkin who was responsible for getting the last main line into London from the north, this being the Great Central Railway's London extension into Marylebone station, opened in 1899. He proposed that this line should continue to the south coast, where a tunnel under the Channel would enable travel from the north of England to France. He was on the board of the French railway company, Chemin de Fer du Nord, which would, he suggested, run the trains from the tunnel on the French side and onward to Paris. He actually set up a company called the Submarine Railway Company to carry out the work and a test bore was made for over a mile under the sea from beneath the cliffs near Dover, to assess the feasibility of the project.

But it wasn't the technical obstacles which defeated his grand plan. The Government's War Office raised the objection that a foreign army could sweep through the tunnel and invade Britain. Despite the nonsense of this argument (what better place to trap an invading army than in the narrow confines of a submarine tunnel), xenophobia won the day and the tunnel scheme was abandoned in the 1890s.

The proposed tunnel terminal in Dover (below).

The mid-channel 'Respiration Station' as conceived in the French plan (below right).

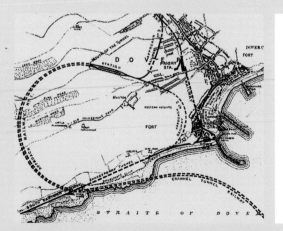

CLASSIC TURN-OF-THE-CENTURY LOCOMOTIVES

Dugald Drummond (1840–1912) was a fine and established locomotive engineer by the time he moved from Scotland, all the way down to the London and South Western Railway (LSWR) in 1895. He had designed successful 4–4–0 locomotives for both his previous employers north of the border, the North British and the Caledonian railways. He immediately set about designing a similar type of locomotive for his new company, but his first attempt, the 'C8', was a flop, being unable to raise enough steam to keep going at the head of express trains. Undeterred, within 12 months he had designed the 'T9' class 4–4–0 and was so confident that it would be a winner that he ordered fifty of them, before even a single prototype was tested: twenty were built at the LSWR's own workshops

at Nine Elms in south London and the remaining thirty were supplied by Henry Dübs & Co of Glasgow. These locomotives were indeed as successful as their designer had predicted and became known to all as 'Greyhounds'. A further sixteen were eventually built and all of them lasted until being taken over by the Southern Region of British Railways in 1948. The last retired in 1961, when they were sixty years old. No. 120 is preserved and is part of the national collection.

Dugald Drummond in a 1900 photograph (left).

The preserved LSWR No. 120 Drummond T9 'Greyhound' hauling a steam special in Surrey, July 1963 (below).

SAFETY AND STANDARDS

In the aftermath of the terrible Armagh crash of 1889, within the year, an Act of Parliament was passed making the interlocking of signals and points, block working and continuous automatic brakes compulsory on passenger trains. This did much to reduce the occurrence of accidents, but it did not eliminate them. There were still two types of error which caused a great many mishaps: signalmen forgetting the presence of a train in their section and drivers running through signals at danger. A third problem was the fitting of gas lighting to carriages, which made the results of an accident all the more lethal should the escaping gas catch fire.

A solution to the first problem had been around since 1872, when American William Robinson invented and patented the 'track circuit'. This was an electrical device which simply detected when a train was on the track, indicating this fact by lighting a bulb on a panel in the signal cabin. However, by 1910 it had only been installed in a handful of locations in Britain, mostly around large terminal stations such as King's Cross.

Inside the cab, the large box houses the solenoids, which operate the valve on the brake pipe. The brass bell is obvious, giving the all-right signal. The warning siren is housed in the small box with a copper pipe to the vacuum system. The driver's cancelling handle is below the siren box.

An early form of ATC, showing the pick-up shoe below the buffer beam of GWR Churchward 2–8–0 No. 4703.

The second problem of drivers ignoring signals was solved by the Great Western Railway with another electrical safety system, which they called 'Automatic Train Control' (ATC). When a locomotive approached a distant signal at danger, both an audible and a visual warning was given in the cab, so, even in foggy conditions, the driver would know that he must slow down in readiness to stop at the next signal. If he failed to acknowledge the warning, the brakes were automatically applied. This system worked superbly and the GWR became the safest railway in Britain. The other companies refused to adopt it, however: a case of 'not invented here'. Very early in the morning of 2 September 1913, a train set off from Carlisle up the long slope to Ais Gill summit on the Settle and Carlisle line

of the Midland Railway. It stalled just before reaching the top. The guard should have protected the train by placing detonators (small explosive devices which make a loud bang when the engine runs over them) on the line behind, but he did not do so. The following train missed all the signals and ran into the back of the stationary carriages. The escaping gas caught fire and sixteen passengers were killed. If ATC had been installed, there is a good chance it would not have happened.

'‘The Countess’ at Castle Caereinon, 1896' (opposite), a painting by Barry Freeman showing a typical turn-of-century scene on a rural branch line.

Shortly after Drummond was perfecting his 4–4–0 for the LSWR, Harry Wainwright (1864–1925) and his chief draftsman, Robert Surtees, were doing the same for the South Eastern and Chatham Railway (SECR) just to the east. The result was, to many people's eyes, one of the most beautiful locomotives ever seen. But the 'D' class wasn't just beautiful. These engines steamed well and rode smoothly and freely, hauling the front rank Kent coast services out of London's Cannon Street station, while receiving admiring attention from fans wherever they went. They didn't look as if they had been put together by an engineer simply as a utility object for lugging carriages along a metal track; they looked more as if they had come from the mind and hands of a craftsman, or even an artist. A total of 51 of these machines were built and thankfully, one has been preserved: No. 737 can be seen at the National Railway Museum in York and will always be affectionately known as Wainwright's 'Coppertop', a reference to the resplendent large, copper dome cover which adorns the top of its boiler.

Another remarkably pretty group of 4–4–0s built at this time, but unfortunately with no surviving members of their class, are the 'Claud Hamiltons' for the Great Eastern Railway (GER). 'Claud Hamilton' was the name of the first member of the class, which first ran in the year 1900 and hence was allocated the number 1900. They were designed by James Holden (1837–1925) and in their royal blue livery with scarlet lining-out, their looks were breathtaking. They worked out of London's Liverpool Street station and pulled expresses to all the major East Anglian towns and to the city of Norwich.

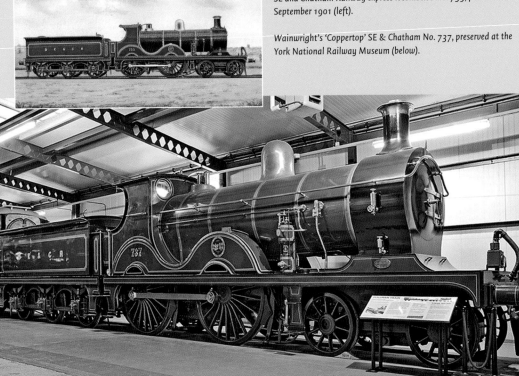

SE and Chatham Railway express locomotive No. 735, pictured in September 1901 (left).

Wainwright's 'Coppertop' SE & Chatham No. 737, preserved at the York National Railway Museum (below).

A GER 'Claud Hamilton' 4–4–0 express locomotive, proudly illustrated shortly after the inauguration of the Norwich Express service in 1900, the year after which the first of the class was numbered.

The 4–4–2 wheel arrangement locomotive first made its appearance in the United States in 1888. Ten years later, the first example of the type was built in Britain for the Great Northern Railway (GNR); it's designer was Henry Ivatt (1851–1923). Ivatt took over as head of the GNR's locomotive department when the venerable Patrick Stirling died in office in 1895. This was not an easy act to follow, but, by the turn of the century, the famous single drivers, epitomized by 'Stirling No. 1', were not finding it easy hauling the top-link expresses out of London's King's Cross station. Something more modern was required, but instead of just building a 4–4–0 like almost every other company was using, Ivatt went for the 4–4–2, the trailing wheels carrying the weight of a larger firebox than could be squeezed between the rear driving wheels of a 4–4–0. The new wheel arrangement was named 'Atlantic', probably because of its trans-Atlantic origins,

An Ivatt C2 'Klondike' 4–4–2 in LNER livery, photographed in the 1930s.

and this first locomotive, No. 990, became the original member of the 'C2' class (strangely enough to be followed by the 'C1' class of Ivatt 'large Atlantics'). The boilers on these locomotives were phenomenal, giving copious volumes of steam, but the cylinders were not able to use all this available performance and the drivers reported that the engines had to be thrashed to get the best from them. When Nigel Gresley took over as Ivatt's successor, he rebuilt the Atlantics with superheaters and larger cylinders. Ironically, when Gresley's own attempts at designing large Pacific locomotives in the 1920s suffered teething troubles, the rebuilt Atlantics often came to the rescue and hauled the express trains. No. 990 is preserved.

George Jackson Churchward was Locomotive Super- intendent and then Chief Mechanical Engineer of the Great Western Railway from 1902 to 1922. His most notable contribution in the early years was the 'taper boiler', the diameter of which reduced from the firebox end of the barrel to the smokebox. This was intended to improve water circulation within the boiler, so that steam would be produced more evenly, preventing pitting on the fire tubes and the firebox walls and using more of the available heat from the furnace. It was first tried out on the 'Atbara' and 'City' class locomotives. Churchward then went on to incorporate it into his later designs of 4-cylinder 4–6–0 locomotives. One of these was the GWR '4000', class or 'Star' class. No. 4003 'Lode Star' is preserved and can be seen at the National Railway Museum in York.

Henry Ivatt in 1897 (above top).

George Jackson Churchward in 1901 (above bottom).

An Ivatt C2 at the head of a GNR express in 1902 (left).

GWR 'Lode Star', designed by Churchward, in the York National Railway Museum (opposite).

THE YEARS OF PRE-EMINENCE, 1906–1919

This was the start of a golden age for the railways, with the network stretching into almost every corner of the country and when motor transport was still in its infancy. Travel became more comfortable for passengers, with the introduction of heated, fully-upholstered coaches with corridors, toilets and dining facilities for longer journeys. Special excursion trains became a common sight, catering for a growing demand from more prosperous, working-class passengers, sometimes taking whole communities on holiday together. Finally, operations during the Great War offered a glimpse of the future, as women temporarily filled the jobs left by men fighting abroad and 130 railway companies were controlled by a single authority.

THE YEARS OF PRE-EMINENCE, 1906–19
THE EDWARDIAN AGE

The years between the start of the new century and the point at which the European nations raced headlong into mutual destruction are often referred to as a golden age, albeit sadly a brief one. This was as true for Britain's railways as it was for any other facet of Edwardian life. At this time, almost all of the country was linked by rail, even the remote parts of Scotland and Wales. Every town either had its own station, or was within a couple of miles of one, while very few villages were more than 16 km (10 miles) away from the freedom offered by the 'iron horse'.

Despite the fact that many of these connections were only branch lines with an infrequent service, in the early 1900s this didn't matter, because the railways had no competition. Country roads were appalling muddy tracks in winter, while in summer they were dusty, hard-baked and rutted. For any journey of more than approximately 16 km (10 miles), the train was the only choice. For shorter journeys, the rich had their horses and carriages, while the poor simply walked. The rapid expansion of the rail network meant that, for the first time, almost everyone could travel almost everywhere.

Movement of goods around the country was also solely by rail. Every station had a goods yard and they were extremely busy places. The railway was what was called a 'common carrier'. This meant that by law it had to transport anything that was requested of it, from an ant circus to an elephant, from bananas to dinosaur dung. This latter may seem strange, but there was a huge industry based on 'coprolite', the technical term for fossilized waste from long-extinct animals: it was used as fertilizer on farms for decades before the chemical industry produced the artificial products that we use now. The passage of goods was meticulously controlled by pieces of paper called 'waybills'. These accompanied the items on their journey, indicating what they consisted of

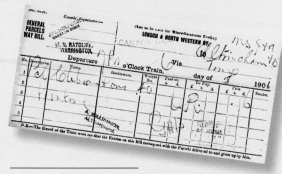

An LNWR waybill from 1906.

and where they should be delivered. Each railway company had a book containing a list of every type of goods or produce which could be imagined. Next to each was the price to be charged for conveying that type of commodity in pence per mile. If, by some unusual chance, someone wanted to send something which wasn't on the list, the station would telephone the head office of the company and they would come back with a price.

Barnstaple Town Station in Devon on a summer's day in 1908 showing the narrow gauge train of the picturesque Lynton and Barnstaple Railway (above).

The crack GWR Irish Boat Express en route to Fishguard, June 1910 (left).

MAXIMUM TRACK MILEAGE

From the 56 km (35 miles) of the Liverpool and Manchester Railway, opened in 1830, the railway network relentlessly expanded throughout the nineteenth century. The way this was done was through raising private capital from speculators – investors eager to make a quick profit. Some of the early routes were obviously going to be financial successes. The main lines centred on London were quickly constructed, the first being the London and Birmingham of 1838. This was already joined to the Liverpool and Manchester by the Grand Junction Railway in 1837, enabling travel from London to either of the two northern cities. The Great Northern line out of King's Cross and the broad gauge Great Western, engineered by Isambard Kingdom Brunel, soon opened for business, along with the sprawl of lines to the south coast towns and resorts, such as Portsmouth, Brighton and Dover.

After these trunk routes had been put in place, investors saw their share values rise and wanted more of the same. Unfortunately, the plum routes were already provided. The 1840s started a rush to build secondary lines of lesser importance, some of which would never make a profit for their owners. This period was later referred to as the 'Railway Mania', when any person could apply for an Act of Parliament to build a line of dubious value, knowing that raising the finance would be easy from gullible investors keen to put their money into the network. At this time, George Hudson (1800–71) rose to prominence as 'The Railway King'. He was a director of the Midland Railway and many other smaller concerns, but later lost all his money and reputation when fraud was revealed, including the bribery of Members of Parliament.

By the end of the 1800s, the extent of the railway network was at its peak of around 34,000 km (21,500 miles) and remained so for the next fifteen years. From the Great War until 1950, line closures reduced the network by an average of 20 km (13 miles) per year. In the brief period from 1950 to 1970, this average was 800 km (500 miles) per year, leaving just half of the total intact.

The southern section of the Railway Map from Baedeker's 1910 guide to Great Britain – the traveller's bible in the early years of the twentieth century. It is fascinating to note that railways, rather than roads, were considered the standard transport arteries of the country until at least mid-century.

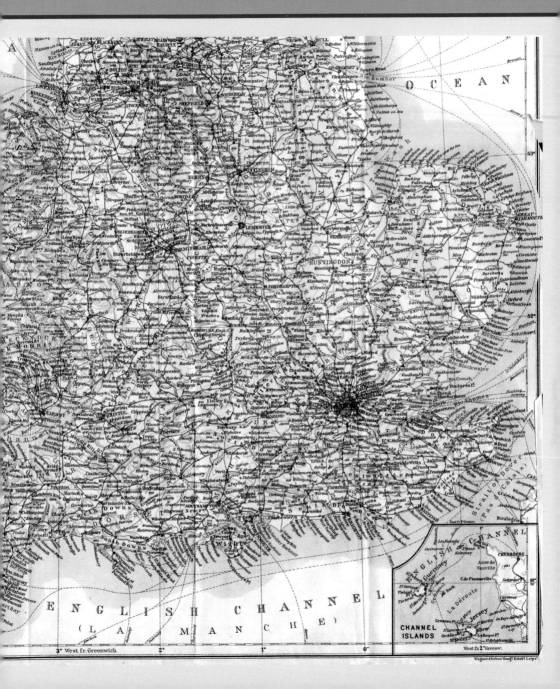

FIRST, SECOND AND THIRD CLASS TRAVEL

British society was strictly divided into social groups and rail travel at the start of the Edwardian era reflected this absolutely. First-class compartments in carriages were the epitome of luxury travel, Edwardian style. The seats were like giant armchairs, sufficient in every respect to accommodate the most corpulent figure. For example, on the Midland Railway, the blue-buttoned cushions were reversible, having deeply piled cloth on one side and smooth leather on the other. Each chair had an adjustable footrest to maintain the toes above the cold floor, which, of course, was sumptuously carpeted. The fittings, such as luggage racks, were of polished brass, ornamented with swirls and flourishes. The doors and window frames were carved and inlaid, while quilted cloth panelling extended a third of the way up the walls. The ceilings had elaborate lincrusta decoration, a deeply embossed covering made from linseed oil and wood flour, set off with a fancy brass surround for the gas, or electric light, of which the latter was becoming more and more common. First-class travel was, of course, used by the country's landowning and historically-wealthy families, but if these had been the only customers, it would not have been economically worthwhile for the railway companies to provide such a standard of travel. The majority of income from first class was provided by the upper-middle classes: doctors, lawyers, stockbrokers, bankers and those in the higher ranks of the military, especially the ubiquitous 'retired colonels' who seemed, if you read the literature of the day, constantly on the move and regular frequenters of first-class compartments.

Second-class was for everyone else who had money or, at least, a sufficient income to provide for their families comfortably, but not extravagantly. A second-class compartment at the start of the twentieth century was not much different from its modern twenty-first century equivalent: cushioned seats and perfectly comfortable surroundings. Second class was most frequented by the many commuters travelling into London for their work every day from counties to the south of London on the London and South Western Railway, which probably accounts for this company being the last to abandon second-class compartments.

Third-class, however, was a different matter. In the early days, the companies did not even provide roofs to the third-class carriages (or trucks, which they more closely resembled). When roofs were eventually provided as standard, the interiors received little in the way of an upgrade. Bare wooden bench seats, rough wooden floors and no lavatories was what you got at the lower end of the social spectrum. Railway companies were forced, by law, to provide travel at a maximum of one penny per mile. When you think that a loaf of bread cost around one penny,

A London, Brighton and South Coast Railway parlour saloon, 1910 (above).

The Great Eastern Region restaurant car service from London Liverpool Street to Norwich, January 1911 (right).

and unsanitary that a third-class carriage would not have seemed a bad place to be, even when undertaking a long journey.

All this was soon to change. It was during the Edwardian period that most companies dispensed with second class altogether, while upgrading third-class to second-class standard, with cushioned seats and toilets. Some companies, especially the Midland, were leaders in bringing more comfortable travel to the masses. James Allport, their General Manager from 1853 to 1880, ensured that the Midland Railway was the first to allow third-class travel on the fastest express trains in upholstered seats with full-height seatbacks. The interior of a third-class Midland Railway compartment was quite luxurious at the turn of the century and far superior to

this actually wasn't very cheap by today's standards, but in the days when the only alternative would have been to walk, it was still a passport to freedom. The train, even in third class, got you from A to B at a reasonable cost and many ordinary people's houses were so uncomfortable

that of any other company. By the end of the Great War, second-class travel had disappeared from the whole of Britain's rail network, with its abolition by the London and South Western Railway in 1918, the last company to take this action.

SUPERHEATING

The superheater, as applied to steam locomotives, was invented by German engineer, Wilhelm Schmidt (1858–1924) in around 1890. Its purpose is to raise the temperature of the steam after it is produced in the boiler. This high-temperature steam is then fed to the valves and the cylinders in the normal way.

The advantage of superheating is that it makes the steam engine more efficient. The second law of thermodynamics predicts that the maximum efficiency with which heat can be turned into mechanical work by a heat engine, of which the steam engine is an example, is determined by the change in temperature from start to finish: the higher the temperature of the steam at the beginning of each piston stroke and the lower its temperature at the end, the more efficient is the engine.

Steam is superheated in a locomotive by collecting it in the usual way at the top of the boiler. It is then passed via a large-diameter steam pipe to the top half of the 'superheater header', which is a collection chamber mounted inside the smokebox. From here, a large number of small-diameter tubes containing the steam pass back and forth the entire length of the boiler within large diameter fire tubes, similar to those which heat the water in the boiler itself. These fire tubes carry hot flue gases from the furnace in the firebox, which heats up the steam in the small tubes. The hot steam is collected in the lower half of the superheater header, then fed via the valves to the cylinders.

A drawing from 1904, showing the location and structure of the superheater in a typical locomotive (above).

North British Railway 4–4–0 No. 11, 'Dominie Sampson', fitted with a Schmidt superheater, January 1918 (below).

The first application of the superheater to a locomotive in Britain was in 1908 on the London, Brighton and South Coast Railway. George Jackson Churchward, of the Great Western Railway, was a great fan of the system and developed his own improved version.

A disadvantage of the superheater is that the valves and pistons run at very high temperatures, so better lubricants are required and much more frequent maintenance. But the reduction in coal consumption of up to 25 percent for the same power output made it worthwhile.

CARRIAGE HEATING

At the beginning of the Edwardian age, on all but the best trains, the passenger had to wrap up warm for the journey: a thick overcoat, a travel rug for the knees and a hand muffler for the ladies. The cold floor made the lower extremities particularly vulnerable, so great use was made of the 'foot warmer', a flat metal tin around 600 mm (2 ft) across containing hot water and sodium acetate crystals. These were available at larger stations, provided on a barrow, which was wheeled alongside the train by a porter, who expected a sixpenny tip from each traveller who took advantage of his service. They were pre-heated when delivered and the recipient placed their feet on top and enjoyed the warmth from below, while it lasted. When they cooled down and thus no longer provided an adequate level of comfort, one could give them a shake, which stirred up the water and the crystals, producing a chemical reaction which heated them again, temporarily.

During the next ten years, steam heating of carriages, provided by a steam pipe running the length of the train and connected to the locomotive's boiler, became more and more common. In 1906, it was still something of a novelty and a selling point to be able to describe railway coaches as being steam heated, but by 1914 almost all main-line passenger trains had been fitted with this comfort-giving device. The small plume of steam, emanating from the leaky valve of the heating pipe on the rearmost carriage of a train, was a common sight on British railways from this time until electric heating was generally introduced in the 1970s and 80s.

A 1908 ladies' magazine pattern for a tapestry cover for your railway carriage foot warmer (above).

ON-BOARD FACILITIES – THE RESTAURANT CAR & WC

When carriages were divided up into individual compartments, each having its own entry/exit door onto the platform, but no connection between itself and the rest of the train. Passengers were confined to this 'cell' for the entire journey, so it was important to use the toilet facilities

G.W.R. Station, Swindon

Swindon station in a 1910 postcard (above).

at the station before embarking. The same went for on-board dining facilities. It was not until the adoption of the 'side corridor' carriage, where the compartments opened onto an internal aisle and the carriages were connected by doors at each end, that toilets and restaurant cars were provided on the train. Before the early 1900s, trains used to stop for twenty-minute comfort breaks at large intermediate stations, such as Swindon on the Great Western Railway. Here passengers could not only use the facilities, but also get some refreshment from the station café and tea-room, at greatly inflated cost for the benefit of the company and its franchise holders.

These new trains with their corridor coaches must have provided relief from anxiety for passengers, especially the ladies, who previously were confined in a compartment with whatever undesirable might chance to step in. Now, if she didn't like the look (or smell – a more common repulsion in those days) of her travelling companions, she could simply escape into the corridor and choose a different seat on the train. In addition, just by walking through the connecting doors until reaching the restaurant car, a good, hot lunch could be obtained for around three shillings and sixpence, thereby avoiding the rush to the station refreshment room during the twenty minutes allowed at York or Swindon. For this sum, the London and North Western Railway's 1906 menu offered: 'Soup, Poached Salmon, Roast Sirloin, Roast Chicken and Salad, Asparagus, Diplomat Pudding, Cheese and Dessert.' Coffee was an extra four pence. The London and North Western Railway was the first to allow third-class passengers to use the dining cars; before this they had been reserved for first-class travellers only.

Some trains had provided lavatories for first-class travellers from the 1880s, but it was not until the advent of the corridor carriage that these could be provided for all. However, many train journeys over shorter distances still did not provide toilets; shops near railway termini offered strange rubber contraptions called 'secret travelling lavatories', which gentlemen could discreetly strap to their leg. Ladies had no such equivalent device and simply had to wait.

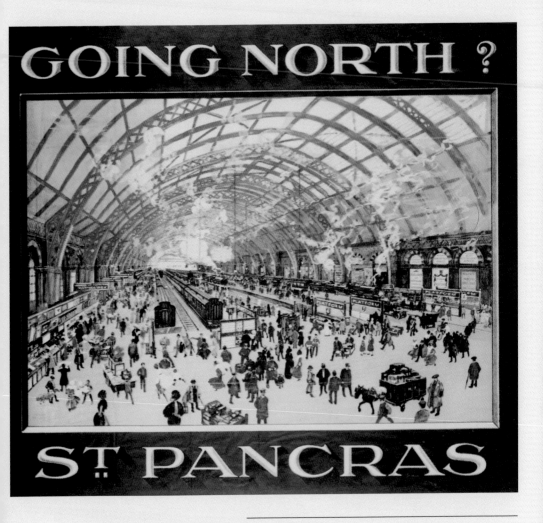

Interior of a North Eastern Region first-class dining car, 1911 (opposite top)

A typical LNER lunch menu from 1905, showing the route on the reverse (opposite bottom).

A Midland Railway poster from 1910, designed by Fred Taylor, showing the bustle of a London main-line terminus (above).

GOODS LOCOMOTIVES

Timothy Hackworth (1786–1850) was the locomotive engineer who, along with William Hedley and Jonathan Forster, was responsible for building the famous, 'Puffing Billy', 'Wylam Dilly' and 'Lady Mary' locomotives for the Wylam colliery wagonway in 1814. On 13 May 1825, he was appointed Locomotive Superintendent of the Stockton and Darlington Railway, four months before the line was opened. The original locomotives, supplied by the Robert Stephenson works, were not very powerful and had problems maintaining boiler pressure. In 1827, Hackworth designed and built 'Royal George'. It had a return-flue boiler, a blastpipe to draw the fire and six wheels coupled together with connecting rods so that all were driven. This was the first example of the 0–6–0 type of locomotive, which went on to become the universal workhorse, either in tender, or tank engine form, for hauling goods trains around Britain for the next 120 years.

At the beginning of the 20th century, there were approximately 20,000 steam locomotives in service on Britain's railways. Over one third of this total was made up of six-coupled 0–6–0 tender locomotives. Their success was due to the fact that all the wheels were driven, thereby giving excellent grip on the rails. They

Timothy Hackworth (left), and the original drawing (below) for his 1827 0–6–0 'Royal George', the prototype for many thousands of goods engines.

Great Northern Region 2–6–0 express goods locomotive No. 1636, pictured in July 1917 (below).

A gallery of tank engines: GWR pannier tank engine No. 3738 (bottom left) at Didcot Railway Centre and 'Nunlow' (top left), a side tank engine on the Keighley and Worth Valley Railway; the saddle tank engine 'Cumbria' on the Furness Railway (top right), and ex-London and South Western Railway 0298 class Beattie well tank engine No. 0314 at the Buckinghamshire Railway Centre (bottom right).

When goods locomotives with more power and grip were required for heavier trains, eight-coupled designs were introduced, either of 0–8–0 type, such as the 30 class, designed by John Aspinall for the Lancashire and Yorkshire Railway in 1900, or the 2–8–0 as introduced on the Great Western Railway by George Jackson Churchward in 1903.

also had small diameter wheels of around 1350 mm (4 ft 6 in) diameter which gave a low gearing, similar to having a car in first or second gear, so that long, heavy trains of loose coupled 10-ton wagons could be pulled up and down hill at speeds which rarely exceeded 48 km (30 miles) per hour.

The final, magnificent example of the British goods steam locomotive was the Robert Riddles designed standard '9F' class 2–10–0 of 1954. A total of 251 were built, the last of which was 'Evening Star', completed in 1960 and the last ever steam engine made by British Railways.

THE SLIP CARRIAGE

The 'slip carriage' was a commonly used device at this time which enabled portions of a train to be directed to different destinations on the line, without the main part having to stop at a station and uncouple. Each slip carriage had its own guard who had control of a lever which, when pulled, disconnected the carriage from its neighbour in front. The brake pipe and steam-heating pipe were sealed by automatically-closing valves so that the main part of the train could continue on its way after disconnection. The slip carriage (or carriages: sometimes there were more than one) became a self-contained train in its own right, under control of the guard, who could bring it to a halt, silently and ghost-like, in the next station.

consisting of just one carriage, others two, or three. The first of these was dropped off at Westbury for onward travel to Weymouth, the second at Taunton, where a carriage for the Minehead branch was slipped and lastly two or three carriages were dropped at Exeter. The main part of the express was, thereby, lightened by four or five carriages for its arduous journey over the steep gradients of the South Devon line between Exeter and Plymouth.

It was obviously important that passengers boarded the correct part of the train at the main station before setting off, otherwise they could end up in the wrong place. The disadvantage of the slip system was that no corridor connection could be made between the slip coach and the one in front, so none of the on-board facilities could be used, such as the dining car. However, the slip coach usually had its own toilet and first- and third-class compartments.

Depending on the speed of the train, it was a skill to release the coupling at the right time, for a fast approach might require release more than a mile before the station, while a slow mover could be slipped just a few hundred yards before it was required to stop. The carriage, on its solitary arrival, would then be hitched to another train that would convey it to its destination, often along a branch line. For example, the Great Western Railway's 'Cornish Riviera Express' often carried three separate slip portions, some

The sensation of arriving at the station and gliding to a halt in almost total silence must have been very strange for those who had not experienced it before.

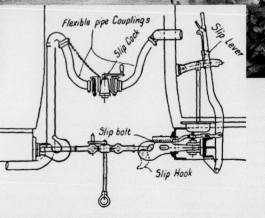

The rear slip carriage being coupled to the 'Cornish Riviera Express' at London Paddington (opposite top).

The up 'Cornish Riviera Express' pictured in November 1912 (opposite).

'Slip', a painting by S.C. Hine (above).

The slip mechanism which allowed the procedure (right).

THE ULTIMATE IN LUXURY AND CONVENIENCE

Looking back, it is astonishing to contemplate the lengths to which the railway companies would go to accommodate the wishes of their richest customers. In Edwardian times, it was possible to charter what was called a 'family saloon' in which the family, its servants and its entire complement of luggage could be transported in a self-contained carriage, like a stately home on wheels. Half of the family saloon would be a parlour, with sofas running lengthways along the walls and separate chairs and tables on which to work or play cards. The next, smaller compartment would be for the servants, fitted with a bell which the masters and mistresses could ring by pulling on a cord in the parlour. The last section of the carriage contained a lavatory and a space for the luggage, of which there was usually a large amount as well-off travellers seemed to take their whole wardrobes and life's other comforts with them.

The family saloon gave exceptionally comfortable and privileged travel for those who had booked in advance, but when necessity meant that a journey had to be made in impetuous haste, an alternative was required. This was to charter a 'special train'. The cost was around seven shillings and sixpence a mile, with a minimum charge of five pounds; to indulge in this extravagance one had to be seriously rich. But if, for example, an American tycoon was too late to catch the boat train from Waterloo to Southampton, hiring a 'special' for a cost of around thirty pounds made good business sense. That the railway companies were able to arrange private travel at such short notice does seem impossible to imagine, but if there was a regular requirement, then some crews and engines were presumably kept on standby for these duties.

Another option was the 'invalid saloon', provided for those with deep pockets and an elderly relative, or sickly child to transport. This was similar to the family saloon, but the patient could be wheeled into the carriage in a bath chair and remain in bed the entire length of the journey, accompanied by other members of the family and attended by servants. Again, the individual service provided to passengers was remarkable, even extending as far as stopping at a station where the train was not booked, just to pick up the patient.

The Bass outing to Blackpool in 1911 (below).

A 1913 North Eastern Railway poster inviting tourists and day trippers to enjoy the bracing pleasures of the east coast seaside resort of Bridlington (opposite).

DAY TRIPS AND EXCURSIONS

At the opposite end of the scale, the working classes were also encouraged to hire their own trains. Since factories had started giving their workers annual holiday leave, there was a boom in works outings to the seaside, especially Blackpool and Scarborough in the north and Brighton and Bournemouth in the south. On Bank Holidays too, the masses flocked onto specially-arranged excursion trains to take them away from the city grime and into the fresh air by the coast.

These 'specials' were sometimes huge, consisting of twenty, even thirty carriages or more. Crammed full of trippers, they could be offered at rock-bottom prices, often just a third of the statutory price of a penny a mile. It was, therefore, possible to travel from Waterloo to Bournemouth and back, a distance of 348 km (216 miles), for just five shillings. Because of the demand and the lower-class nature of the travellers, the rolling stock used on these trains was usually old and inferior, being kept in dusty sidings until required for each summer season. Many of the carriages were the old unheated and unpadded four- and six- wheelers, which had been built in mid-Victorian times. But everyone was out for a good time and, with such low prices, no one really cared as long as they got their day at the seaside. Some of the large employers, for example the Bass brewery in Burton-on-Trent, would charter a dozen or more trains specially to take all their staff on the annual outing to Blackpool, leaving in the early hours of the morning and returning late at night. In 1909, the Great Western Railway offered a Bank Holiday special day trip from Paddington to Newton Abbot non-stop, a distance of 309 km (192 miles).

RAILWAYS DURING THE GREAT WAR

In 1912, the possibility of war had been anticipated with the formation of a Railway Executive Committee which would run the railways as one system if the need arose. Soon after Britain entered the war on 4 August 1914, the railways were quickly providing extraordinary support for the armed forces. In just eight days in August 1914, 69,000 men, 22,000 horses and 2,500 guns, with their associated baggage, fodder and ammunition were transported by rail to the port of Southampton and thence to northern France.

At the start of hostilities there was not one ambulance train in existence; by the end of August, twelve of these had been converted from normal rolling stock and were ready for service. Lord Kitchener, the commander-in-chief, said that the railways had performed every task that had been requested of them and that all units had arrived in France 'well within the scheduled time'.

The aftermath of the Quintinshill disaster (left top)

Troops heading for Dover and the French front, 1915 (left bottom).

The Railway Executive Committee had 34,329 km (21,331 miles) of track under its control and 130 railway companies. There were around 600,000 staff employed by the railways at the outset, but 180,000 of them immediately joined up. Staff shortages meant that women were given jobs working on the railways.

Some railway companies especially had difficulty coping with extra wartime traffic. The British Grand Fleet was based at Scapa Flow, off the north coast of Scotland. It required prodigious quantities of coal to feed the ships' boilers and the Highland Railway had to borrow locomotives from other companies to fulfil the demand.

The Caledonian Railway also provided a good deal of the extra coal traffic and had difficulties of its own, the most glaring of which was the Quintinshill disaster of 22 May 1915, when 226 people, mostly soldiers of the 7th Battalion, The Royal Scots, on their way to Gallipoli, lost their lives in Britain's worst ever train crash.

Requests to the general public to travel less during the war generally went unheeded. Many trains lost their restaurant cars. While cheap fares were abolished, ordinary fares increased by 50 percent and services were cut. Government control of the railways did not finish with the Armistice of 11 November 1918. Power was not returned to the individual companies until 15 August 1921.

One of dozens of railway company war memorials, this is the Cambrian Railway's memorial at Oswestry in Shropshire (right).

GWR 2–6–0 No. 5322, which was built in October 1917 and sent to France, along with other members of the class, to haul supply trains between Calais and the front line railheads. The locomotive is now preserved at Didcot Railway Centre (below).

THE ZENITH OF STEAM, 1920–1946

The efficiencies produced by running the railways in a coordinated fashion during World War I demonstrated the advantages of amalgamation. The Groupings Act of 1921 brought most of the country's 178 railway companies under the control of the 'Big Four': the LMS, LNER, GWR and SR. Their collective investments in engineering created some extraordinary locomotives and led to iconic individual achievements, in particular those associated with the second great race to the north between the LMS and LNER. But the days of the Big Four were numbered. The railways were hit by economic depression and competition from road transport and they were stretched to breaking point by the demands placed upon them during World War II.

'Days of Red and Gold – Lune Valley, October 1939', painted by Barry Freeman. LMS 'Coronation' class 4–6–2 No. 6239 'City of Chester' powers northwards with an express for Glasgow. 'City of Chester' was one of the last locomotives to appear in red and gold livery, entering service shortly after the outbreak of World War II. It was painted black in 1943 and de-streamlined in 1947.

THE ZENITH OF STEAM, 1920–46

By the end of the Great War in 1918, there were 178 separate independent railway companies. During the war the government had appointed a Railway Executive Committee to oversee railways for the duration. Effectively under state control, the railways had operated efficiently, pooling locomotives and rolling stock. Peacetime brought with it claims for compensation from the companies and also an increasing belief that rationalization, even nationalization, could produce a better railway. The Railways Act of 1921, sometimes called the Groupings Act, made provision for the companies to be amalgamated and absorbed into just four major companies. If the companies failed to agree merger terms, an Amalgamation Board was empowered to force them to do so.

This time nationalization was avoided and the railways were grouped into what became known as the 'Big Four' companies: The Great Western Railway (GWR), London and North Eastern Railway (LNER), London, Midland and Scottish Railway (LMS) and the Southern Railway (SR).

The four new railway companies inherited a mixed bag of locomotives and types, many of which had been built before the turn of the century and had performed throughout the Great War with the minimum of maintenance. For example, at its inception, the LNER inherited 7,700 locomotives, 20,000 coaches, nearly 30,000 freight vehicles, some 120 items of electric rolling stock, plus steam ships, eight canals, twenty docks and harbours and 23 hotels. The Big Four urgently needed new locomotives.

THE GWR'S CASTLES AND KINGS

In 1922, Charles Collett (1871–1952) became Chief Mechanical Engineer of the Great Western Railway and set about updating existing 'Star' class 4–6–0 four-cylinder engines to produce a more powerful locomotive. The result was the 'Castle' class, which made its first appearance in 1924. The locomotives featured a longer boiler, enlarged firegrate area and larger cylinders, plus a roomy cab with extended roof and seats for the crew. The low-sided tender carried six tons of coal and 3,500 gallons (around 16,000 litres) of water (increased to 4,000 gallons in 1926). The 'Castle' class locos were, at that time, the most economical in the country and, measured by tractive effort (pulling power), the most powerful, too, at 141 kN (31,635 lb), all this accomplished within the axle-weight restrictions of the GWR. The copper-capped chimney and polished brass safety valve completed the elegant look of these fine locomotives. By the time production ended in 1950, 171 'Castles' had been built.

LMS expresses passing at speed – facing the artist is No. 6100 'Royal Scot', January 1934 (opposite).

GWR down West of England express, headed by locomotive No. 4099, 'Kilgerran Castle', July 1927 (below).

When the 'Castles' were introduced, the GWR publicity department took great delight in advertising them as 'the most powerful locomotives in Britain'. However, when the Southern Railway introduced its 'Lord Nelson' class in 1926, they were no longer able to claim this title and Collett was asked to produce a locomotive with a target tractive effort greater than 40,000 lbs. He achieved this with the 'King' class of 1927, but it is arguable whether this was really a better and more powerful locomotive than the 'Castle' class, or rather more of a publicity stunt on the part of the Great Western Railway. By fitting a larger boiler with a 25 psi increase in pressure and 6 mm (¼ inch) greater diameter cylinders, the engine was almost 10 tons heavier than a 'Castle' and therefore exceeded the permitted axle loading for the track – and it still didn't give the desired 40,000 lbs of tractive effort. The increase in tractive effort was accomplished by fitting 60 mm (2½ in) smaller diameter driving wheels, thereby reducing the gearing to give the required figure. The whole thing was a little contrived and, although these were immaculate engines, they necessitated the strengthening of many miles of track on the GWR to withstand their extra weight. As a result, they were confined to the main-line routes between London and Plymouth in the West Country and London and Wolverhampton in the Midlands.

The first 'King', No. 6000 'King George V', completed in June 1927, was shipped to America to participate in the centenary celebration of the Baltimore and Ohio Railway. A total of thirty 'Kings' were constructed, 29 of them surviving until 1962, when they were withdrawn from service.

CASTLES, KINGS AND HALLS

Despite the depression that followed World War I, the Great Western Railway experienced a significant increase in holiday traffic to the West Country resorts in Devon and Cornwall. The GWR's 'Star' class 4–6–0 locomotives (not to be confused with the earlier broad gauge 'Star' class) had performed well since their introduction in 1905 by the GWR's Chief Mechanical Engineer, George Jackson Churchward, but a more powerful locomotive was required to pull heavy holiday trains. Charles Collett's solution was the four-cylinder 4–6–0 'Castle' class, based on the 'Stars', but including a new boiler operating at increased pressure, producing greater tractive effort. A total of 171 'Castles' were built at Swindon Works between 1923 and 1950. The locomotives were noted for their superb performance and fuel economy.

The 'Castles' were joined in 1927 by the 'King' class 4–6–0 locomotives, the largest locos the GWR ever built. Named after monarchs of the United Kingdom, the 'Kings', designed by Collett, were intended to cut the journey time between Paddington and Plymouth to just four hours. The heavier weight of these engines, at 136 tons including the tender, meant that they were restricted to the GWR's main-line routes from London to Penzance, Birmingham, south Wales and Chester and some cross-country routes between the Midlands and the west. Between 1927 and 1930, thirty 'Kings' were built.

GWR 'Cornish Riviera Express', hauled by No. 4074 'Caldicot Castle', pictured in January 1925.

The 'Halls' were a class of two-cylinder 4–6–0 locomotives and they introduced the concept of a 'mixed traffic' type (capable of hauling both passenger trains and freight trains). The initial batch of fourteen locomotives, sent for proving trials on the Cornish main line, were an instant success. Between 1928 and 1943, 259 of them were built at Swindon to Collett's design; a further 71 'modified Halls', incorporating improvements designed by Collett's successor, Frederick Hawksworth, were built between 1944 and 1950. All carried names of famous houses in the GWR's area.

GWR No. 7029 'Clun Castle', No. 4936 'Kinlet Hall', and No. 4965 'Rood Ashton Hall', photographed at Tyseley, Birmingham, June 2008 (below).

GWR No. 5900 'Hinderton Hall', photographed at Didcot Railway Centre, May 2010.

THE BIG FOUR

In the 1840s, during the period referred to as the 'railway mania', hundreds of tiny railway companies were formed, eager to persuade shareholders to put their money into lines which ran from 'not a lot' to 'nowhere in particular'. Some built not one inch of track, failing at the first hurdle of raising capital or gaining the required Act of Parliament, while others built their line, but then immediately went bankrupt. The larger companies saw these failures as opportunities. Most of the cost of a railway was in the civil engineering required to build the trackbed. If a new venture company performed this task and then had insufficient funds to run the train service, it was forced to sell its assets, often at a price far below that of the cost of constructing the line. Soon, a few 'super companies' emerged, each dominating its own geographical area of the country.

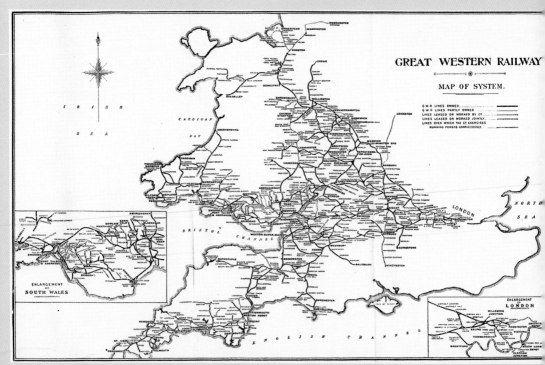

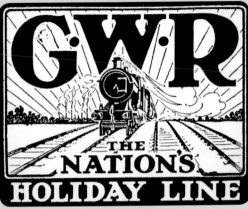

The GWR 'Cornish Riviera Express' hauled by 4–6–0 No. 6018, 'King Henry VI', July 1934 (opposite top).

The Great Western Railway (GWR) commanded its area of the West Country and south Wales and was a prime member of the 'Big Four' British railway companies.

The railways to the south of London were many and varied, running to all the south coast seaside resorts and ports, as well as the north and east Kent coasts. At the 1923 grouping, these were all bundled together to become the Southern Railway (SR), another constituent of the Big Four.

A 1923 map of the Great Western network (opposite).

The Southern Railway 'Pullman Continental Express', hauled by King Arthur class 4–6–0 No. 770, 'Sir Prianus', July 1926 (above left).

The GWR promoted itself as 'The Nation's Holiday Line', as in this poster from the 1920s (above).

A Southern Railway network map from 1936 (below).

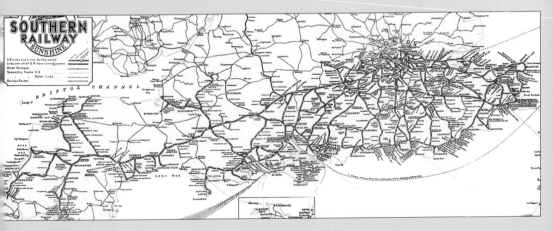

The other two members of the Big Four were formed from the lines running north out of London, all the way up into Scotland.

The East Coast Main Line ran from King's Cross via York, Berwick-upon-Tweed and Edinburgh. It had originally been formed from the Great Northern Railway, the North Eastern Railway and the North British Railway. At the grouping, these became part of the London and North Eastern Railway (LNER).

An LNER up east coast express, with 4–4–2 No. 9902 'Highland Chief' at its head, January 1927 (below).

The Forth Bridge was one of the LNER's big attractions, as this 1928 poster by H.G. Gawthorn underlines (bottom).

The Southern Railway was renowned for the ultra-modern design of its posters; among its designers were H. Molenaar (above) and T.D. Kerr (below).

The last of the Big Four was made from a quite unhappy alliance between the London and North Western Railway (LNWR), which ran from London's Euston north to Birmingham, Crewe and Carlisle and the Midland Railway (MR), which left St. Pancras on its way past the headquarters at Derby and then onward over the famous Settle and Carlisle line. These two former rivals were forced to merge into the London, Midland and Scottish Railway (LMS).

A 1935 LMS network map (below).

LMS 4–6–2 No. 6203 'Princess Margaret Rose' heads a train on the Carlisle–Settle line, January 1937 (top right).

A stylish 1924 LMS poster by Norman Howard (right).

GRESLEY'S PACIFICS

The LNER was fortunate in having the services of Hebert Nigel Gresley (1876–1941) as its Chief Mechanical Engineer. Gresley had already held responsible positions with the Lancashire and Yorkshire Railway and the Great Northern Railway, where he succeeded Henry Ivatt as Chief Mechanical Engineer in 1911. He took up his post with the LNER on 1 January 1923, the day it was formed.

Gresley's long and successful connection with Doncaster Locomotive Works began in 1911, when it built his powerful K1 and K2 class goods locomotives (later adapted for passenger work). Gresley's designs were truly innovative, exemplified by his 'conjugated valve gear' to control the central cylinder on three-cylinder locomotives, patented in 1916 and his patented double-swing suspension system for pony-truck wheels (a 'pony truck' is a single-axle truck fitted at the front or rear of a locomotive to help the engine negotiate curves – a two-axle truck is called a 'bogie').

A Gresley K1, LNER No. 2005, photographed in 1965 (opposite).

LNER up non-stop 'Flying Scotsman' 4–6–2 No. 4476 'Royal Lancer', pictured in January 1929 (above).

Gresley's most famous locomotives were his 4–6–2 A1/A3 and streamlined A4 Pacifics. The first of these entered service for the Great Northern in 1922 and following tests in 1923, the A1 was chosen by the fledgling LNER as its standard, main-line locomotive.

In 1924, The LNER's No. 4472 'Flying Scotsman' was exhibited at the British Empire Exhibition, alongside the first member of the GWR's 'Castle' class, No. 4073 'Caerphilly Castle'. The following year, the two railway companies agreed to an exchange of locomotives for comparative trials. The GWR's 4079 'Pendennis Castle' spent some months with the LNER, where it proved to be the equal of Gresley's larger A1 Pacifics in haulage capacity and high-speed running, but did so much more economically, using 10 percent less coal while achieving similar, or better performance. Gresley took good note of these results; he soon modified the valve gear on his Pacifics to improve performance and economy by making fuller use of the steam before it was exhausted and fitted boilers which ran at a higher pressure. These improved locomotives were designated the A3 class; the A1s received the same upgrades when their own boilers became due for renewal.

By the 1930s, the railways were beginning to suffer from competition from motor cars, long distance coach services and new commercial domestic airlines. Gresley realized that the railways would need to compete with faster trains capable of speeds of more than 160 km (100 miles) per hour. Gresley undertook experiments, during which time A1 4472 'Flying Scotsman' exceeded 160 km (100 miles) per hour and A3 'Papyrus' set a speed record of 174 km (108 miles) per hour.

The LNER board granted Gresley permission to proceed with his A4 'Silver Jubilee' streamlined trains. The stream-lining extended beyond the locomotive to include the train's carriages. In a demonstration run on 27 September, 1935, one train exceeded 180 km (112 miles) per hour. The service from King's Cross to Newcastle commenced four days later, completing the non-stop 435 km (270 mile) run in just four hours. Further A4 streamlined locomotives and trains were built to extend the service to Edinburgh and to add a new service to Leeds and Bradford.

GRESLEY AND RATIONALIZATION

At the 1923 Grouping, the LNER was formed through merging seven major railways. The LNER also had interests in six joint railway companies and absorbed, or managed a further four smaller companies. All these brought with them their own engine policies and requirements, from heavily-used suburban commuter lines to lightly-used rural branch lines and including millions of tons of coal, ore and freight traffic across the industrial north of England.

With many different types of locomotives in service, a good number of them elderly, the benefits of standardizing on fewer types, using common parts where practical, were obvious. As well as designing his A1/ A3 and A4 Pacifics for main-line passenger work, Nigel Gresley was instrumental in producing standard, efficient and practical engines for lighter work, mixed traffic and heavy-freight trains. The first of these were the 'J38' and 'J39' classes, versatile 0–6–0 two-cylinder locomotives, designed at the LNER's works at Darlington. In 1926, thirty-five 'J38s' were built for the LNER's freight operations in Scotland. The 'J39' was a modification of the 'J38', featuring larger wheels. In a fifteen-year period, two hundred and eighty-nine 'J39s' were built. Two of the 'J38s', 65901 and 65929, became the last two Gresley locomotives in British Railways service, when they were finally withdrawn in 1967 after forty years' service.

A Gresley J38, British Railways No. 65917, photographed at Dunfermline in 1965 (opposite).

The only surviving K4 locomotive, BR No. 61994 'The Great Marquess', now based at Thornton Junction (above).

Gresley had designed his 'K' class 2–6–0 'Mogul' freight locomotives for the Great Northern Railway. Gresley looked again at the design when a need for a powerful locomotive arose for the remote West Highland Line to Fort William in Scotland. The first 'K4' used cylinders from the 'K3' design and boiler from the 'K2' design and was introduced in 1937, with five more following in 1938. All six were adapted to take snowploughs.

Gresley's efforts then produced his V1/V3 2–6–2 tank engines and the V2 2–6–2 tender engines. The three-cylinder V2 'Green Arrow' class was one of his most successful types, with a total of 184 being built between 1936 and 1944. Intended for express goods work, many of these mixed traffic engines saw passenger service, especially during World War II.

WILLIAM STANIER AND THE LMS FIGHTBACK

Because of the Midland Railway's 'small engine' policy, by the 1920s, it was lagging behind its rivals in terms of motive power, still double heading most of its expresses (using two locomotives at the front of the train). Gresley's success in adapting GWR design and practices for his A3 locomotive probably influenced the London Midland and Scottish Railway, formed at the grouping in 1923, to ask in 1926 if they too might run trials with a GWR 'Castle' between London and Carlisle. The GWR's No. 5000 'Launceston Castle' performed well – so well that the LMS asked if the GWR might build a batch for the west coast line. When the GWR directors turned this down, the LMS then asked if it could have a set of construction drawings and permission to build 'Castles' under licence. This too was refused, so Henry Fowler introduced 'The Royal Scot', a 4–6–0 express passenger locomotive, based on the Southern Railway's 'Lord Nelson' class. A total of seventy of these three-cylinder locomotives were built between 1927 and 1930 and went some way to providing the increase in power that was required.

On 1 January 1932, a new Chief Mechanical Engineer was appointed at Derby. William Stanier (1876–1965), previously Works Manager at Swindon, brought with him the secrets of how to build successful large locomotives like the ones on the Great Western Railway. With the backing of the chairman of the LMS, Stanier reversed the old Midland Railway's 'small engine' policy and was responsible for the introduction of four famous classes of main-line locomotives: the 'Black Fives', the 'Jubilees', the 'Princess Royals' and the 'Princess Coronations'.

The 'Black Fives' were introduced in 1934 and 842 of them had been built when production ceased in 1951 (production was suspended during World War II). These incredibly successful 'mixed traffic' 4–6–0 engines with two cylinders used a variation of the GWR's tapered boiler. They could do just about anything asked of them, which is why so many were made and were still in service, right up to the end of steam with British Railways.

Coronation class No. 46229 'Duchess of Hamilton' in unstreamlined condition at Tyseley, May 2006.

Construction of the 'Jubilees' lasted from 1934 to 1936 and a total of 191 were built. Made specifically for main-line passenger work, these 4–6–0 three-cylinder engines were often to be found on LMS lines between Derby and London. Initially, they were not good steamers, but modifications to the super-heaters turned them into very good locomotives.

The LMS Crewe Works then built thirteen 'Princess Royal' Pacifics between 1933 and 1935. These were four-cylinder 4–6–2 locomotives for express passenger trains, especially the 'Royal Scot' between Euston and Glasgow.

The 'Princess Coronation' class 4–6–2 locomotive was an enlarged version of the 'Princess Royal' and was so named to commemorate the coronation of George VI. Produced by the LMS to compete with Gresley's A4s on the LNER, it proved to be the most powerful steam locomotive ever to run in Britain, as measured in trials and much more powerful than any of the diesel engines that replaced it. The 'Art Deco' movement was in full swing in the mid-1930s and streamlining on cars and other vehicles was the height of fashion. The first five 'Coronations' were produced with a curvaceous blue-painted casing around the engines, lined out in silver, which looked very stylish. Matching carriages were added for the 'Coronation Scot' train, which was reproduced in stylized form on countless posters. The next five examples of these engines had their streamlined casings painted in the traditional 'Crimson Lake' colour of the Midland/LMS railway (see pp102–103). Lined out in gold, they arguably looked even better than the blue ones. A total of 38 'Princess Coronations' were built between 1937 and 1948, some of which were produced in unstreamlined form, but even during the war some were still made

with the streamlined casings fitted, albeit painted black. The streamlining on all the locomotives was eventually removed for ease of maintenance.

Gresley's achievements on the east coast prompted another railway 'race to the north', this time between the LNER and the London, Midland and Scottish Railway. In July 1937, Stanier and the LMS responded to the LNER's east coast successes by launching the 'Coronation Scot'. This named train service to Scotland was hauled by the 'Princess Coronation' class Pacific locomotives, which could maintain a six-hour schedule from London to Glasgow. On a run from Euston in 1937, No. 6220 'Coronation' set a new British speed record of 184 km (114 miles) per hour south of Crewe, taking the record from the LNER. In doing so, it entered the pointwork at Crewe too quickly and, although luckily staying on the rails, crockery was smashed in the dining car as it lurched from one side to the other. A year later, on a test run between Glasgow and Crewe, No. 6234 'Duchess of Abercorn', hauling a rake of twenty coaches giving a total load of 605 tons, recorded a power output of 2,483 kW (3,330 horsepower) in the cylinders, the highest of any British steam locomotive before, or since.

Gresley, though, was not to be outdone. On 3 July 1938, in great secrecy and masquerading as a 'brake test', streamlined A4 Pacific No. 4468 'Mallard' regained the world speed record for a passenger steam train for the LNER, reaching 203 km (126 miles) per hour, a record which has never been beaten.

Former LMS Stanier 5P5F 4–6–0 'Black Five' No.5407 at Waterscale between High Bentham and Clapham Junction, working the Carnforth to Hellifield leg of a Cumbrian Mountain Pullman charter, November 1982.

INCREASING POWER ON THE SOUTHERN

The Southern Railway consisted of lines between London and the south coast and London and the west of England, the latter competing with the GWR. The Southern pursued a policy of electrifying its largely commuter services to south coast destinations, but needed suitable locomotives, principally for services to Dover and to the west beyond Salisbury. The first of the Southern Railway's designed-and-built locomotives was the 'Lord Nelson' class, which took the GWR's outright power record away from them. Richard Maunsell (1868–1944) had been Chief Mechanical Engineer (CME) of the South Eastern and Chatham Railway from 1913 and on the Grouping in 1923, he became the SR's CME. The 'Lord Nelson' class was the SR's four-cylinder 4–6–0 locomotive, initially intended for express passenger trains from London to Dover, but later also working express trains to the west of England. Introduced in 1926, sixteen of them were built and remained in service until 1962.

The 'Lord Nelson' class worked alongside the Southern Railway's 'King Arthur' class. These 4–6–0 two-cylinder engines, first built by the London and South Western Railway in 1919, were continued by the SR until production ceased in 1926, by which time a total of 74 had been constructed.

SR's proud announcement of 'Lord Nelson' in 1926 (right).

Restored SR No. 850 'Lord Nelson' at Minehead in September 2006 (below).

SR 'Merchant Navy' class No. 35017 'Belgian Marine' leaves
Waterloo at the head of the 'Bournemouth Belle', 1950 (above).

Restored SR No. 21C123 'Blackmore Vale' at Sheffield Park on
the Bluebell Line, July 2006 (right).

In 1937, Oliver Bulleid (1882–1970) succeeded Maunsell
as Chief Mechanical Engineer at the Southern Railway.
Bulleid had joined the Great Northern Railway at
Doncaster in 1901 and later worked for Westinghouse
and the Board of Trade. In 1912, he rejoined the Great
Northern working for Nigel Gresley. After his Great War
service, he returned to the GNR and, in 1923, Gresley
brought him back to Doncaster to be his assistant. With
this extensive experience, Bulleid designed the 'Merchant
Navy' class of 4–6–2 'air smoothed' streamlined Pacifics
for the Southern Railway, the first of which, 'Channel
Packet', was built in 1941, followed by 29 others up
to 1949. The class was among the first to use welded
construction, enabling easier fabrication during World
War II and featured unique chain-driven valve gear, the
chain being enclosed in a sealed oil bath.

After the war, the Southern Railway introduced its 'West
Country' and 'Battle of Britain' classes of 'light' Pacifics.
A total of 110 were constructed, again with 'air smoothed'
streamlining. Being lighter in weight than their 'Merchant
Navy' counterparts, they were able to operate on a wider
variety of routes, often appearing in such far-flung places
as Barnstaple and Ilfracombe in Devon.

TRACTIVE EFFORT – MEASURING THE MIGHT OF THE STEAM LOCOMOTIVE

The tractive effort of a steam locomotive tells us how big a load it can pull and this benchmark has been used since the earliest times to measure an engine's might. It was often in the past and sometimes still is, referred to as the locomotive's 'power', although actually it is no such thing: it just tells us the force with which the pistons can push the wheels around from a standing start. To measure power (in the proper sense), the engine has to be moving and a specialist piece of equipment called a 'dynamometer car' was added to the train in order to carry out this measurement, the power actually varying at different speeds and with different load conditions. Tractive effort can be calculated simply from knowledge of the diameter and stroke of the pistons, the number of cylinders, the diameter of the driving wheels and the boiler pressure.

The formula for the calculation of tractive effort is:

$$TE = \frac{N \times D^2 \times S \times 0.85P}{2W}$$

- TE is the tractive effort in pounds
- N is the number of cylinders
- D is the diameter of the pistons in inches
- S is the stroke of the pistons in inches
- P is the boiler pressure in pounds per square inch
- W is the diameter of the driving wheels in inches

The convention is that tractive effort is calculated assuming that 85 percent of the maximum available boiler pressure is applied to the pistons, hence the 0.85P in the formula.

To appreciate steam locomotive progress, look at the following tractive effort values for various locomotives:

- Stephenson's 'Rocket' (1829): 820 pounds
- GWR broad gauge 'Iron Duke' class (1847): 8,262 pounds
- GNR Stirling Single (1870): 11,130 pounds
- Midland 4–4–0 compound (1902): 23,205 pounds
- Southern Railway 'Lord Nelson' class (1926): 33,510 pounds
- GWR 'King' class (1927): 40,300 pounds

GWR No. 6024 'King Edward 1' at Wantage, December 2009.

SR No. 850 'Lord Nelson' at York, December 1982 (opposite).

THE AGE OF RECORDS

During the 1930s, the 'Big Four' companies promised speed and style, no doubt encouraged by their respective publicity departments, anxious for attention-grabbing headlines. Speed was of the essence. The first to claim it had the fastest regular train in the world was the Great Western Railway, with its Cheltenham Spa Express, nicknamed the 'Cheltenham Flyer'. Even before the Great War, the train was scheduled to cover the 146 km (91 miles) from Kemble Junction, Gloucestershire, to Paddington in 103 minutes, an average speed of over 85 km (53 miles) per hour. The introduction of Collett's 'Castle' class locomotives enabled the speed to be accelerated and on 6 June 1932, between Paddington and Swindon, 5006 'Tregenna Castle' set an average speed record for the train of 131.3 km (81.6 miles) per hour. That September, the regular service was scheduled to complete the Swindon to Paddington run at more than 114 km (71 miles) per hour.

The Southern Railway was not to be outdone. In 1926, it was able to claim to have the most powerful locomotive in Britain when the first of its 'Lord Nelson' class locomotives was introduced. These sixteen engines were originally intended for use on the Southern's 'Golden Arrow' luxury boat train, providing a first-class only Pullman service between London and Paris, via Dover.

The publicists were busy again when Gresley's streamlined Pacific locomotives and trains enabled non-stop services between London and Newcastle in 1935, running at a top speed of 145 km (90 miles) per hour. Clearly, fast trains between London and Scotland would be attractive and in 1937, the London, Midland and Scottish Railway began its 'Coronation Scot' service between London and Glasgow on the west coast main line, using streamlined engines and setting a new steam locomotive speed record of 184 km (114 miles) per hour on 28 June. Two days later, the LNER attempted to beat this on a demonstration run, but only managed 175 km (109 miles) per hour. But on 3 July 1938, the LNER's A4 Pacific No. 4468 'Mallard' set the world speed record for steam locomotives at 203 km (126 miles) per hour on the slight downward gradient of Stoke Bank, south of Grantham on the East Coast Main Line.

LNER No. 4490 'Empire of India' hauling the 'Coronation Express', January 1938 (below).

Terence Cuneo's famous 1980 painting of BR No. 60022 (formerly LNER No. 4468) 'Mallard' (right).

'The Cheltenham Flyer' in a photograph of August 1923 (above).

GWR were keen to promote the service, and produced a special book (right) 'for boys of all ages'.

THE RAILWAYS IN WARTIME

At the outbreak of World War II, the railways once again found themselves under government control through another Railway Executive Committee. Locomotive construction slowed, as works around the country turned to supporting the war effort. However, construction of some heavy-freight locomotives was authorized and some express locomotives were reclassified as 'mixed traffic' locos, suitable for both passenger and freight work.

During the war, the government's Directorate of Transport Equipment was led by Robert A. Riddles (1892–1983). Riddles had worked with Stanier at the LMS and had been responsible for much of the design work on the 'Coronation' class Pacifics. At the Directorate, Riddles designed three locomotives suitable for wartime service. These 'Austerity' designs were cheap and easy to build and maintain and could use poor quality coal, or even oil, as fuel. The 0–6–0 saddle tank locomotive, built by Hunslet of Leeds, went on to become a post-war 'standard' locomotive for use at collieries and other industrial sites. Riddles adopted a Stanier designed '8F' 2–8–0 as a standard military goods engine, before designing his own 2–8–0 locomotive, based on the Stanier '8F' and using many inter-changeable parts. Between 1943 and 1945, 935 of these were built.

Riddles then designed a 2–10–0 locomotive based on his 2–8–0 design. Between 1943 and 1945, 150 of them were built and shared many common parts with the 2–8–0.

The war department had little work for these locomotives before the Normandy landings, so many of them were 'run in' by the LNER and the LMS on their services, before being shipped off to Europe in 1944. After the war, a good number of the 2–8–0 'Austerities' saw further service with the LNER and remained in use almost until the end of the British Railways steam era.

The Southern Railway found itself thrust into the frontline from the onset of hostilities, many of its routes being close to the English Channel. The company lacked suitable freight locomotives for the extra war work, so the CME, Oliver Bulleid, quickly designed a stripped-down Q1 0–6–0 engine, which in 1942 became the most powerful 0–6–0 ever to run on Britain's railways. In 1942, forty of these locomotives were built. Although intended as freight engines, they could also be used on passenger trains. All of them remained in service until the 1960s.

Wartime austerity did not end with hostilities in 1945. The railway companies, starved of investment in rolling stock and locomotives and operating a network that had been kept going with the minimum of maintenance, faced a time of shortages, not just in materials, but of skilled labour and manufacturing capability. Their future was decided with the election of a government that had committed itself to nationalization.

A restored Bulleid Q1, No. C1, photographed at a York Railfest (opposite).

'Evacuees leave Paddington, 1940', a painting by Stella Whatley (above). The railways were crucial to the evacuation effort – in just three days at the beginning of September 1939, 3.5 million people were evacuated from Britain's cities.

RAILWAYS DURING WORLD WAR II

World War II provided the railways with a different set of challenges from the Great War. Aerial bombardment was a serious threat and enemy action was aimed at destroying the country's ability to move wartime traffic. At the declaration of hostilities, the railways performed the incredible feat of evacuating 1.3 million children from the major cities to safer country towns in nearly 4,000 special trains. Troop trains were soon carrying servicemen to the ports at Glasgow and Southampton. The retreat to Dunkirk saw the railways move 300,000 survivors in ten days in May and June 1940 in two hundred trains.

Domestic trains were severely reduced and journey times extended. Trains carrying troops and supplies were given priority, causing normal services to be held up. Materials to repair bomb damage to tracks and stations were in short supply and the poor state of the track led to speed restrictions being imposed.

Wartime also presented safety problems. Platform edges were painted white to make them easier to find in the blackout. In November 1940, the blackout contributed to an accident at Norton Fitzwarren in Somerset, when a driver misread semaphore signals on leaving Taunton and his train derailed, killing 27 people.

The railways played an important part in the defence of Britain. The north–south lines provided a barrier to

The aftermath of the Norton Fitzwarren accident (left).

SR locomotives built just before the war for the Dunkirk boat route (shown are Nos. 1758 and 1145) were critical to the movement of troops in the early days of the war (opposite below).

'Wartime Goods', painted by Ray Schofield. The Staines Branch, shown here, was a vital link for the conveyance of armaments to the south coast (below).

potential invasion and hundreds of 'pillboxes', defensible gun positions, were built alongside railways. Holes were drilled in strategic railway bridges to enable them to be destroyed to attempt to halt any invasion. Construction of hundreds of airfields, mainly in the east of England, relied on the railways to transport materials, supplies and personnel. Some rail-mounted anti-aircraft guns, hauled by dedicated locomotives, were constructed and placed strategically in Kent. The military even commandeered the narrow gauge Romney, Hythe and Dymchurch Railway, mounting guns and armour plating on the tiny trains. Preparations for the D-Day landings in 1944 saw the railways running five hundred special trains every day.

THE TWILIGHT OF STEAM

The 'Big Four' railway companies became one organization, British Railways, following nationalization in 1948, but politicians' pledges to invest in the badly battered railways would not be fulfilled. Within twenty-five years, Britain's railway network had shrivelled to half of its previous size, as the impact of increased car ownership, management blunders and political underhandedness led to wave after wave of line closures. The manufacture and mainline deployment of steam locomotives came to an end after one final flourish of engineering brilliance. Yet this was one of the most fascinating periods in the history of Britain's railways, the consequences of which would lead directly to a renaissance of steam during the final quarter of the century.

'The Rare Bird — Peterborough, 1951', painted by Barry Freeman. The appearance of one of Nigel Gresley's A4 'Pacifics' always caused excitement amongst lineside observers. No. 60024 'Kingfisher', an Edinburgh-based locomotive seldom seen so far south, appears out of the gloom of a winter's day with an express bound for King's Cross.

THE TWILIGHT OF STEAM

At the end of World War II, Britain's railways were in a pitiful condition. Falling profits during the years of the Depression had restricted attempts by the Big Four companies to modernize their networks before the war, while the strains of the conflict itself had exacted a heavy toll on rolling stock, track and buildings. But the experience of running the railways as a united entity under Government control during the war had demonstrated the potential advantages to be gained from such an arrangement.

Following a landslide victory in the 1945 General Election, the Labour Government was committed to nationalization of Britain's primary industries and utilities, including the railways. The 1947 Transport Act, intended to form the foundation for a fully modernized, integrated transport strategy, came into force on 1 January 1948, creating British Railways, which would be controlled by the new British Transport Commission (BTC).

The Big Four and fifty-five smaller railway companies were now merged into a single organization, leaving only a handful of narrow gauge lines in private hands. The new entity was divided into six operating regions. The Great Western and the Southern remained largely intact, becoming, respectively, the 'Western Region' and the 'Southern Region' of British Railways. The new 'London Midland Region' covered most of the old LMS area without Scotland. The most affected of the Big Four was the LNER, which not only lost all its Scottish lines, but was further divided into the 'Eastern Region' and the 'North Eastern Region' of British Railways, a seemingly perverse decision (they were reunited again in 1967). All the Scottish routes became part of a new 'Scottish Region'.

The cover and map from British Railways: The New Organisation, *outlining the future of railways in Britain.*

A 1950 BR poster in the spirit of the age by Terence Cuneo, showing a busy scene at Paddington (below left).

FORGING AHEAD

THE 1947 TRANSPORT ACT AND NATIONALIZATION

Despite glorious feats of speed by individual locomotives and the development of high-profile luxury services during the 1930s, the Big Four railway companies all suffered financially during the Depression, while freight revenues were also undermined by an expansion in road haulage. By 1939, they were relying on ageing rolling stock and infrastructure, which were then requisitioned by Government and worked to breaking point in support of the war effort. By the end of the war, the system was in urgent need of renewal.

Clement Attlee's post-war Labour government created the British Transport Commission (BTC) as part of its nationalization programme. The BTC was to oversee railways, canals and road freight transport across Great Britain. Under the 1947 Transport Act it had a duty to provide 'an efficient adequate, economical and properly integrated system of public inland transport'. British Railways, the name given to the Railway Executive of the British Transport Commission (BTC), assumed control of the Big Four and of fifty-five other smaller railways not already incorporated in the 1923 amalgamations. The new company began the task of building a financially and operationally stable organization, fit for a country reconstructing itself in the aftermath of war. Unfortunately, not only did the Government never allocate the resources required to repair and renew the track and rolling stock, it imposed conditions on the newly nationalized system which meant that it had to 'buy itself' from its previous owners, the private railway company shareholders. The railways generated 70 percent of the BTC's total income but only some of this revenue was reinvested in the railways, with the rest diverted by the BTC to fund improvements to canals, docks, harbours and road transport.

Pages from the 1947 Transport Act (right), and the BR 'cycling lion' logo, used from 1950 to 1956 (below).

ENGINEERING TALENT

While there were inevitably some political and bureaucratic disputes as the managers and employees of some very different companies began to learn how to work together, British Railways could now benefit from the fruits of their combined engineering talent.

Oliver Bulleid (1882–1970), Chief Mechanical Engineer (CME) at the Southern Railway, had produced some fine new locomotive designs during the war, most notably a three-cylinder 4–6–2 Pacific built in two classes, 'Merchant Navy' and 'West Country' (some of which were later renamed as the 'Battle of Britain' class). Following nationalization he became CME for BR's Southern Region and produced further steam and diesel loco prototypes.

H. G. 'George' Ivatt (1886–1976) became CME of the LMS in 1945 and went on to oversee production of LMS-designed locomotives during the early years of British Railways, including 'Royal Scot' and 'Patriot' class locomotives and more of William Stanier's 4–6–0 'Black Five' engines. The first two main-line diesel-electric locomotives, numbers 10000 and 10001, were also built at Derby during this period.

The great Sir Nigel Gresley (1876–1941) was succeeded as CME of the LNER by Edward Thompson (1881–1954), whose successes included the B1 class 4–6–0 mixed traffic locomotive and the introduction of steel-bodied coaches, which were much safer than Gresley's beautiful, but weaker teak-bodied stock. They were the direct ancestors of the British Railways Mark 1 coaches, some of which were still in service at the start of the twenty-first century.

Thompson's successor, Arthur Peppercorn (1889–1951) became CME in 1946, then served as CME for BR's Eastern and North Eastern regions until 1949. He left his mark in the shape of the reliable 'Peppercorn' class A1 4–6–2 locomotives that hauled express trains on the East Coast main line between London, North East England, Edinburgh and Aberdeen from 1949 until 1961, some of the most efficient and reliable steam locomotives ever built. None was saved for preservation, but in 2008, the A1 Steam Locomotive Trust completed the construction of 60163 'Tornado', built using Peppercorn's blueprint.

Finally, British Railways' Western Region would inherit the great GWR Works at Swindon, where the standard of precision engineering was arguably higher than at any of the other companies' works. Following nationalization, Swindon came under the command of Alfred Smeddle (1899–1964), an engineer trained by the LNER at Darlington, who modified many old 'Castle' and 'King' class GWR locomotives in the early 1950s, adding double-blast chimneys and new superheaters to give them one last lease of life before retirement.

BR 'Battle of Britain' class No. 34059, 'Sir Archibald Sinclair', preparing to leave Sheffield Park station on the Bluebell Line, November 2009.

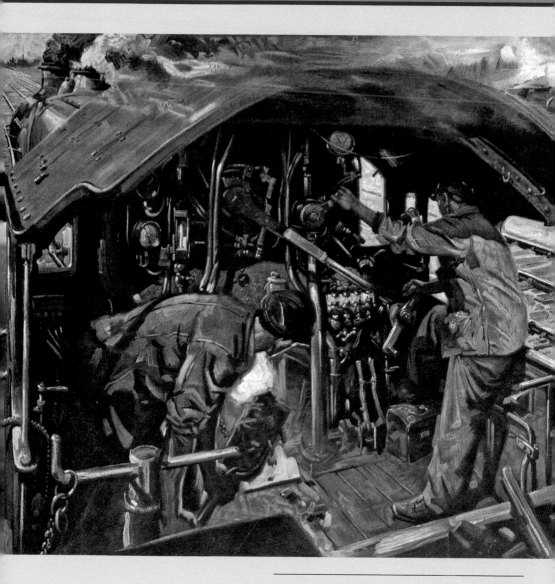

Another optimistic British Railways poster by Terence Cuneo – painted in 1949 and entitled 'Full steam ahead', it shows the cab of ex-GWR 'Monmouth Castle', speeding betwen Paddington and Reading at over 70 miles per hour.

ONE LAST BLAST FOR STEAM

But if British Railways inherited some fine locomotives and talented engineers, the rest of the new company's assets and infrastructure were more of a mixed blessing. The Big Four bequeathed 20,000 locomotives to BR, 40 percent of which were already at least 35 years old in 1948, as was 20 percent of the coaching stock. There had been major bomb damage to stations, bridges and tunnels during the war and BR was faced with a backlog of routine track maintenance equating to 2,500 track miles, according to the official British Railways history of the period (Gourvish's, *British Railways 1948–1973*). The Chief Inspecting Officer of the Railways blamed this lack of maintenance for a post-war spike in the number of accidents on the railways in his annual report of 1947.

These difficult conditions encouraged a series of compromises. The most significant was that British Railways backed away from large-scale investment in electrification, or dieselization, instead commissioning hundreds of new steam locomotives. This was good news for steam locomotive drivers and firemen and for railway enthusiasts, but is now viewed by some railway historians as a contributory factor in the large number of line closures during the 1950s and 1960s. More extensive use of diesel multiple units (DMUs) during the 1950s might have generated sufficient efficiency savings to save many minor and secondary routes.

But in the late 1940s, the BTC Railway Executive took the view that the supply of coal was more reliable and likely to remain lower than oil prices. So, under the leadership of R. A. 'Robin' Riddles (1892–1983), Member of the Railway Executive for Mechanical and Electrical Engineering, and his two principal assistants, Roland C. Bond and E. S. Cox, British Railways began to build twelve new Standard locomotive classes designed to meet the operational needs of the railways as efficiently as possible, until such time as it were economically and practically possible to consider the use of diesel or electrification. They included some outstanding locomotives, such as the 'Britannia' and 'Clan' passenger classes and the celebrated '9F' 2–10–0 class, including 'Evening Star', the last steam locomotive built for British Railways, completed in 1960.

Coping with war and its aftermath was a major headache for the newly-formed British Railways – bomb damage had created a massive capital and maintenance deficit. Shown here is bomb damage at Derby, Coventry, Nuneaton and London St Pancras.

THE DRIVE TO MODERNIZE

Standardization and the improved reliability and efficiency of locomotives and other rolling stock helped to improve operational and financial performance, but the railways also faced challenges related to wider social changes. One theme of the 1950s was a growing public awareness of the negative effects of industrial pollution, highlighted by episodes such as the Great London Smog of 1952, which led to four thousand deaths across the city. A growing public appetite for cleaner modern technologies did not sit well with dirty, smoky steam locomotives. Meanwhile, road transport continued to take freight business away from the railways throughout the 1950s and private car ownership was increasing quickly. The railways were losing the pre-eminent position in public and freight transport that they had enjoyed for more than a hundred years.

British Railways' response to this gathering crisis was its Modernization Plan, implementation of which began in 1955, designed to win back passengers and freight by improving the reliability, safety, speed and capacity of the railways. Its principal measures included electrification of main lines in the Eastern Region, Kent, the Midlands and central Scotland, replacement of steam with diesel power across the network; new rolling stock, track and signalling, and closure of lines deemed uneconomical or to be wastefully duplicating existing routes.

The Plan led to some useful changes in the longer term, in particular the electrification of suburban services around London and Glasgow and – over the course of the next twenty-five years – electrification of both the West Coast and East Coast main lines and of various other parts of the network. But in the shorter term, it was an expensive disaster, costing £1 billion. There was heavy investment in freight marshalling yards and new shunting locomotives, rather than any meaningful attempt to address the fundamental reasons why more goods were being transported by road. Each of the operating regions competed for investment without adequate supervision, or strategic direction from the centre. Dieselization was rushed through too quickly, with some inadequately-tested diesels proving so unreliable that they had to be withdrawn from service within a few months of their introduction. At the same time, the phasing out of steam locomotives also accelerated rapidly, meaning some lines were, temporarily, very badly under-resourced. It was a public relations disaster that failed in its primary aims.

SAFETY IMPROVEMENTS AND THE RISE OF THE TRAINSPOTTER

There were brighter moments for the railways in the 1950s. Safety improvements were pushed forward following the Harrow and Wealdstone disaster in 1952, when a crash involving three different trains left 112 people dead, and another terrible crash at Lewisham in south London in 1957, when an express train collided with a local service in fog, killing ninety people. In each case, the accident was caused by a driver missing a red signal, so could have been prevented by the Automatic Warning System (AWS), an alert system which forces drivers to take action when approaching a signal set at red or yellow, the original version of which was implemented by the Great Western Railway as long ago as 1906, making it the safest railway system in the country. British Railways began to install AWS across its network from 1956, but its use did not become mandatory until the late 1990s.

The 1950s was a strange period in railway history: the end of an era. It's not surprising that this was the decade when 'trainspotting' became a popular pastime. True, that was a reflection in part of the restricted leisure choices available to schoolchildren at the time, but there was also a remarkable range of rolling stock to spot. British Railways was introducing its Standard locomotive classes, while stock inherited from the Big Four included many locomotives, coaches and wagons dating back to the 1920s and earlier. Former LMS and LNER Pacifics, for example, might be seen side-by-side with a Black 5 LMS locomotive, former Midland Railway '4P' class compound locomotives, or the veterans of the GWR's 'Castle' and 'King' classes.

The Harrow and Wealdstone disaster of 8 October 1952, Britain's worst peacetime railway accident.

THE GREAT SURVIVORS

An amazing number of tank engines continued to showcase the brilliance of Victorian engineering, including the 'A1' and 'A1X' class 0–6–0T 'Terrier' tanks, built at Brighton works in the 1870s for the London, Brighton and South Coast Railway. A compact design with a low axle load that somehow packed enough power to pull twelve-coach trains, it was used for decades to haul London commuter trains over lines built with shallow ballasting, tight corners and steep gradients. In their early years, these trains ran through Marc Brunel's original Thames Tunnel between Rotherhithe and Wapping, after it was incorporated into the East London Railway network in 1865. Others went on to be used on various railways around the world as well as on the Isle of Wight and the Hayling Island branch in Hampshire, with the last of them not withdrawn from service until 1963. You can still see two of them at the Isle of Wight steam railway today.

Other tank survivors included the Midland Railways 1377 class 1F 0–6–0T, eleven of which, built at Derby and Newton-le-Willows in the 1880s, were still shunting at Staveley ironworks in Derbyshire in the early 1960s; the London and North-Western Railway's 'Coal Tank' 2F 0–6–2T, built at Crewe between 1881 and 1897, which hauled coal, but also passengers in Wales and North West England in the 1950s; and the London and South Western Railway's 415 class 1P 4–4–2T, built in Manchester, Glasgow and Newcastle in the 1880s and still hauling passenger trains on the Axminster to Lyme Regis branch in Dorset until 1961.

Among larger engines to survive into the post-war era were eleven of the Holmes North British Railway's C class 0–6–0 six-coupled goods engines, built at Glasgow between 1888 and 1900, two of which (65288, built in 1897, and 65345, dating from 1900) were the last working steam engines still in regular service in Scotland in 1967; the Lancashire and Yorkshire Railway's 27 class 3F 0–6–0 freight locomotives, built at Horwich Works between 1889 and 1918, fifty of which were still in use by British Railways at the end of the 1950s; and GWR 2800 class 8F 2–8–0 locomotives, built at Swindon between 1903 and 1919, which were only superseded by the appearance of the formidable BR Standard 9F class in the 1950s.

Trainspotting in London in the 1950s (opposite).

Terrier tank engine 'Freshwater' on the Isle of Wight Steam Railway, May 2006 (below left).

Perhaps a secondary attraction for the trainspotters was that it was clear from the middle of the 1950s that the door was starting to close on the steam era. There was already poignancy in the sight of older locomotives reduced to less glamorous duties, such as GWR 'Castle' class locomotives hauling local passenger services on various parts of the network, usually looking far more dirty and neglected than would ever have seemed possible in their pre-war heyday. Their sad condition foretold a worse fate in store for most steam locomotives.

Preserved LSWR radial tank 4–4–2T 415 class No. 288 (right)

South Eastern and Chatham Railway C class No. 592 running on the Bluebell Railway (below).

GWR 'Hall' class No. 4936 'Kinlet Hall' with 'The Lincolnshire Poacher' at Attenborough, July 2009.

A NEW GENERATION OF STEAM LOCOMOTIVES

The man who oversaw the creation of the last generation of British steam locomotives was Robert Arthur 'Robin' Riddles (1892–1983). He began his career as an apprentice at Crewe Works for the LNWR in 1909 and rose to hold senior engineering roles at the LMS in the 1930s. Following work for the Ministry of Supply during World War II, when he designed the WD 'Austerity' 2–8–0 and 2–10–0 freight locomotives, Riddles served as Vice-President of the LMS from 1944, until his appointment as Member of the Railway Executive (the body that oversaw the creation of British Railways) for Mechanical and Electrical Engineering, in 1947.

British Railways had inherited 20,000 steam locomotives from the Big Four railway companies, a collection that varied widely in age and quality. More than 8,000 were already at least 35 years old. With large-scale conversion to electric or diesel traction economically unthinkable in the near-term, new steam locomotives were needed to meet the needs of the railway. Riddles and his two principal assistants, Roland C. Bond and E. S. Cox, oversaw the design of twelve new Standard locomotive classes. In engineering terms, many were very successful, but their working lives were destined to be short.

Class 8 Pacific 'Duke of Gloucester' heads 'The Mid-day Scot', August 2006.

PASSENGER EXPRESS LOCOMOTIVES

Class 8: 4–6–2 Pacific
'Duke of Gloucester' class 71000

Only one class 8 was ever built, to replace the 'Princess Royal' class locomotive 46202 'Princess Anne', destroyed in the Harrow and Wealdstone railway crash in 1952. 'Duke of Gloucester', built at Crewe in 1954, was based on the Standard class 7 Britannia design and incorporated Caprotti valve gear. It performed inconsistently and was removed from service in 1962 after just eight years of service. After initially becoming part of the National Railway Collection, then falling out of favour again, it seemed certain to be broken up, but was bought in 1974 by a group of enthusiasts, who spent thirteen years restoring it and correcting its design flaws. It is now frequently in steam at locations across the UK.

Class 7: 4–6–2 70000 to 70054
'Britannia' class

Britannia locomotives were designed for high-speed passenger services and fifty-five were built at Crewe between 1951 and 1954. They were generally popular with crew and used throughout the network until the late 1960s. Two survive: 70000 'Britannia' and 70013 'Oliver Cromwell', which hauled the very last BR steam passenger service, from Liverpool to Carlisle, in August 1968.

Class 6: 4–6–2 72000 to 72009
'Clan' class

The Clans were Pacific 4–6–2 locomotives, smaller than the Britannia class, with a modified boiler, smaller cylinders and other weight-reducing features, enabling their use on a wider variety of routes. Only ten were built, between 1951 and 1952, and none were preserved. But a new Clan is being built, 72010 'Hengist', on the proceeds of public donations and sponsorship, by the Standard Steam Locomotive Company, based in Devon.

The new 'Clan' 6MT class locomotive, 'Hengist', currently being built in Devon (above).

Former British Railways Standard 7MT 4–6–2 'Britannia' class No. 70013 'Oliver Cromwell' and Standard 8P 4–6–2

FREIGHT AND MIXED TRAFFIC LOCOMOTIVES

Heavy Freight Locomotives

Class 9F: 2–10–0, 92000 to 92250

In the 1950s, 250 of these magnificent 2–10–0 locomotives were built at Crewe and Swindon, with the 251st and last, 92220 'Evening Star' (the only named '9F'), completed at Swindon in 1960. The '9F' was designed to pull freight trains of up to 900 tons at 56 km (35 miles) per hour with maximum fuel efficiency, covering long distances within a footplate crew's eight-hour shift, thus generating invaluable cost savings. They also pulled many passenger services. Nine of this class have been preserved, including 92203 (named since preservation 'Black Prince'), which set a record for the heaviest train ever hauled by a steam locomotive in Britain in 1983, hauling 2,162 tons at a quarry in Somerset.

Mixed Traffic Locomotives

Class 5: 4–6–0, 73000 to 73154

This was a class of 172 locomotives based on the successful LMS Stanier class 5 ('Black Five') 4–6–0 and built at Derby and Doncaster between 1951 and 1957. Of these, thirty were built with Caprotti valves, including 73129, one of five saved for preservation.

Class 4: 4–6–0, 75000 to 75079

Locomotives from this class were designed for duties unsuitable for the heavier class 5 and eighty were built at Swindon (1951–57). Those sent to the Southern region were given high-sided 4,725 gallon tenders to make up for the lack of track water troughs. Six are in preservation.

Class 4: 2–6–0, 76000 to 76114

This was a versatile mixed traffic locomotive and 115 were built at Doncaster, Horwich and Derby between 1952 and 1957. Four survive.

Class 3: 2–6–0, 77000 to 77019

Only twenty of this small mixed traffic engine were built at Swindon in 1954. None survive.

Class 2: 2–6–o 7, 78000 to 78064

These engines, known as 'Mickey Mouse' engines, were the smallest of the standard classes. They were built at Darlington between 1952 and 1956. Four have survived into preservation, with one, 78059, now being rebuilt as a 2–6–2T, the tank version of the class, at the Bluebell Railway in Sussex.

BR Standard class 9F No. 92203 'Black Prince' (left).

BR Standard class 4 No. 80002 on the Keighley and Worth Valley Railway (below).

Passenger Tank Locomotives

Class 4: 2–6–4T, 80000 to 80154

Designed for suburban rail services, these locomotives were used on lines such as the London, Tilbury and Southend. Between 1951 and 1956, 155 were built at Brighton, Derby and Doncaster. Of these, fifteen have been preserved.

Class 3: 2–6–2T, 82000 to 82044

None of these class 3s survive, although forty-five were built at Swindon between 1952 and 1955.

Class 2: 2–6–2T, 84000 to 84029

Between 1953 and 1957, thirty of these tank engines were built at Crewe and Darlington. All were scrapped, but a former class 2 2–6–0 78059 is to be rebuilt as a 2–6–2T.

BR Standard class 2 No. 78019 at Leicester North on the Great Central Railway (below).

MODERNIZATION AND THE BEGINNING OF THE END FOR STEAM

The need to modernize what was still, in many respects, a Victorian railway network and the potential efficiency gains connected with the use of either electric or diesel traction instead of steam, had been recognized long before World War II. But the prospect of more widespread use of electric or diesel traction still seemed distant in the late 1940s, when an economic crisis and assumptions about fuel supplies and costs made it natural to assume that steam power would continue to dominate the railways in the short to medium term. So, British Railways began building thousands of new steam locomotives during the early 1950s – only to reverse the policy suddenly in 1955, following publication of the British Railways Modernization Plan.

The Plan's measures included electrification of several main lines in England and Scotland and replacement of steam with diesel power across the network. In the longer term, some of these changes would have positive effects. In the short term, however, the Plan was an expensive disaster. Millions were wasted on new freight rolling stock and marshalling yards, with the changes made too slowly to counter the trend of freight moving to the roads. The worst example of waste was, perhaps, the Bletchley flyover on the West Coast main line, built at a cost then of £1.6 million as part of a proposed, but never completed, scheme to turn the Oxford to Cambridge line into a useful east–west route that avoided London. The line was closed in the 1960s and the flyover was never used.

Dieselization was also rushed through too quickly and some inadequately-tested diesel locomotives were so unreliable that they had to be withdrawn from service within a few months of their introduction. Nonetheless, the Plan marked the beginning of the end for the use of steam locomotives on British Railways.

In 1958 Woodham Brothers' scrapyard at Barry in South Wales won a contract from British Railways to scrap locomotives and by the mid-1970s, when the photograph opposite was taken, more than two hundred locomotives remained at the yard, which became a mecca for railway preservation enthusiasts (opposite).

The infamous Bletchley Flyover, photographed in the late 1970s (below).

'EVENING STAR'

British Railways Standard class '9F' 92220 'Evening Star' was the last steam locomotive built by British Railways, completed at Swindon in 1960. It is the physical embodiment of the end of an era and was also the 999th BR Standard class locomotive to be built. It was earmarked for preservation before it had even entered service.

'Evening Star' was the only one of the '9F' heavy freight class to be given a name when constructed. It was also given a copper-capped double chimney and painted in Brunswick Green livery, which was usually reserved for passenger locomotives. Although 'Evening Star' and a number of other '9F' class locomotives did haul passenger services at various times and locations, all the other '9F' locomotives were painted black. Its number is not the highest of the '9F' class, as some locomotives with later numbers had already been completed. The last of the '9F' class in numerical order, 92250, was also the last steam engine to be built for British Railways at the Crewe Works, in 1958.

'Evening Star' being prepared for the naming ceremony at Swindon Works on 18 March 1960 (left).

Hauling the 'Cumbrian Mountain Express' on the Settle–Carlisle line, April 1984 (below).

A WORKING LIFE CURTAILED

The name 'Evening Star' was chosen by three British Railways employees, joint winners of a competition run in the BR Western Region staff magazine: Driver T. M. Phillips, Boilermaker J. S. Sathi and one F. L. Pugh.

The locomotive had a very short working life of just five years, mostly spent on the Western Region and working on the hilly Somerset and Dorset line, where it hauled the last ever 'Pines Express' in September 1962. It was withdrawn from service in 1965 and became part of the National Railway Collection. The engine's permanent home is the National Railway Museum in York, although it is steamed regularly and has also spent time at other railway museums, including STEAM, the GWR museum housed in the former works at Swindon, where the locomotive was built.

'Evening Star's nameplate, showing the special plaque inscribed 'No. 92220 built at Swindon March 1960, the last steam locomotive for British Railways. Named at Swindon on March 18th 1960 by KWC Grand, Esq, Member of the British Transport Commission' (left).

'Evening Star' hauling the 'Scarborough Spa Express' out of York, August 1983 (below).

THE RESHAPING OF BRITISH RAILWAYS

In the late 1950s, conversion to electric and diesel traction and the phasing out of steam was accelerated across the network. At the same time, many of the lines on which they had hauled trains also began to disappear. Dr Richard Beeching (1913–85) is usually seen as the chief culprit for the destruction of so much of Britain's railway heritage in the 1960s, but the process was already well underway in the preceding decade, when the BTC's Branch Line Committee closed two hundred loss-making lines across the country.

This still left BR with 17,500 track miles, 6,800 stations, 600 marshalling yards and ongoing financial problems. The company made a loss of £15.6 million in 1956; by 1960 this had grown to £42 million. The railways had few friends in Government. Ernest Marples (1907–78), transport minister in Harold Macmillan's Conservative government after the 1959 General Election, was also co-owner of a road-building company, Marples, Ridgeway & Partners. Marples did not hold shares in his company while in office: they had been sold, with a purchaser's requirement attached, that Marples should buy the shares back after he had left office at the original price, if the purchaser so required. The purchaser turned out to be his wife.

In fact, British Railways' passenger business was not in bad shape; revenues had increased by £2 million in 1959. It was the decline of freight business, in which so much had been unwisely invested following the Modernization Plan, that was dragging the railways down. But when Marples appointed Richard Beeching as head of the new British Railways Board, which replaced the BTC (and so swept away any semblance of an integrated transport strategy), little attention was paid to such subtleties. Beeching's brief was simply to bring the railways back into profit.

Beeching's report, *The Reshaping of Britain's Railways*, was published in 1963. Its headline findings were damning, in black and white economic terms: a quarter of fare income came from 34 stations, while half of the rest of the network contributed just 2 percent of income and only 4 percent of parcels business. Beeching recommended the end of many stopping services, to be replaced with new bus services and closure of more than a third of the network's stations and 5,000 miles of 'uneconomic' lines.

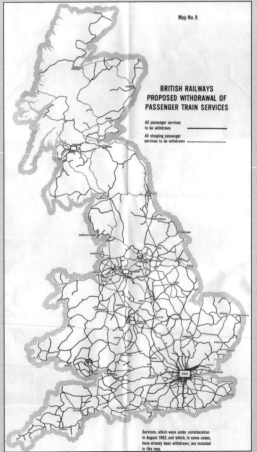

Map No.9

BRITISH RAILWAYS PROPOSED WITHDRAWAL OF PASSENGER TRAIN SERVICES

All passenger services to be withdrawn ————

All stopping passenger services to be withdrawn ··········

Services, which were under consideration in August 1962 and which, in some cases, have already been withdrawn, are included in this map.

His research had not taken into account the wider social value of the railways, or the consequences of increased road traffic. It looked at the railways in isolation, rather than examining the economics of other forms of transport. The figures upon which assessments of individual lines were based had been collected during April, thus ignoring the rise in holiday traffic on many lines during the summer months, which would have provided a quite different picture of their true economic worth.

In 1964, the Conservatives lost the General Election, but the Labour Government went back on a pledge to oppose Beeching's cuts. Instead, the rate of closures accelerated. By the end of the 1960s, almost half of the stations in operation at the start of the decade had closed, along with more than 6,000 miles of railway, including much of the Great Central main line to the Midlands from London Marylebone, the beautiful Somerset and Dorset line, and the Oxford to Cambridge line, a useful east–west link avoiding London.

Ernest Marples (opposite left) and Richard Beeching (above), the architects of the sweeping railway closures during the 1960s.

The key plan of closures from Beeching's 1963 The Reshaping of Britain's Railways (opposite right).

THE END OF STEAM
– AND A NEW BEGINNING

Together, these changes helped ensure the final demise of steam. Between 1952 and 1968, the last year of routine steam services on Britain's railways, more than 20,000 steam locomotives were withdrawn and scrapped. A few were retained for specific purposes, tanks held in reserve for shunting coal trains when needed, but most of these were withdrawn over the next decade.

Yet the story of the railways in the 1950s and 1960s was not entirely bleak. For one thing, we can be thankful that the destruction of steam locomotives on such a huge scale was not followed by the destruction of all Britain's railway heritage. This was an era when British Rail (as it became in 1964) routinely vandalized its assets in the name of progress, but its cavalier approach played a major role in boosting the profile of conservation organizations, like the Victorian Society. For example, the demolition of the Victorian London Euston station in 1961–62 in order to construct a new replacement helped galvanize opposition to threatened demolitions elsewhere, including St Pancras

station, a few hundred yards away, along Euston Road. That this stunning building, perhaps the greatest of London's Victorian railway stations, would survive to enjoy a new life in the twenty-first century, as the refurbished terminus for high speed train services to continental Europe via the Channel Tunnel, is a direct result of a movement set in motion by the destruction of the railways in the 1960s. It is just one of dozens of beautiful main-line and branch-station buildings renovated over the past twenty-five years.

Finally, while it is easy to regret the withdrawal of steam and the loss of so many railway lines that could have been of great practical benefit to the country, as well as of huge historical interest, the events of these decades also inspired and made possible the development of Britain's uniquely rich collection of preserved steam railways, where steam locomotives still reign supreme, waiting to be discovered and enjoyed by new generations of admirers.

'TORNADO' –
THE TRIUMPHANT RETURN OF STEAM

'Peppercorn' class A1 60163 'Tornado' is a miraculous delight: a brand new passenger express steam locomotive, built to a sixty-year-old design.

The original 'Peppercorn' class A1 4–6–2 was a three-cylinder, locomotive designed for the LNER under the direction of Chief Mechanical Engineer, Arthur Peppercorn. Forty-nine were built at Doncaster and Darlington in 1948–49, to haul fifteen-coach passenger trains on the former LNER main line between London and Scotland.

The locomotives were reliable, as well as strong. By 1961, they had racked up more than 77 million km (48 million miles), equivalent to 325 km (202 miles) per locomotive per day, and could travel up to 190,000 km (120,000 miles) between each works overhaul. This stamina, in combination with large grates that allowed the useof lower grade coal, made the A1 extremely cost-efficient. Yet they were all withdrawn and scrapped between 1962 and 1966.

Remarkably, that was not the end of the story. In 1990, the A1 Steam Locomotive Trust began raising money to build a new A1. Donations from thousands of members of the public and financial support from companies including Rolls-Royce, Corus and BAE Systems, meant construction could begin. Eighteen years later, the Trust had raised almost £3 million and was able to complete the construction of 'Tornado', named in honour of the pilots who flew Tornado fighter jets during the first Gulf War of 1991.

'Tornado' crosses Bawtry Viaduct, south of Doncaster, working 1Z63 London Kings Cross–Edinburgh Waverley, filming for BBC's 'Top Gear' programme, April 2009.

In January 2009, the engine hauled her first main-line passenger service, from York to Newcastle. 'Tornado' has a theoretical top speed of 160 km (100 miles) per hour and can haul trains of ten or eleven coaches. The original design has been modified to add water capacity and electronic safety equipment, but in all other respects, seeing 'Tornado' is like watching a ghost restored to life. When, in 2008, the locomotive moved under its own steam for the first time at Darlington, Arthur Peppercorn's widow, Mrs Dorothy Mather, rode on the footplate. 'My husband would be very proud,' she told the press.

'Tornado' on the turntable at York Railway Museum (left).

'Tornado' passes through Flax Bourton on its way to Plymouth, hauling the 'Tamar Tornado', August 2009 (below).

Three ex-British Railways locomotives at Bo'ness in June 2009. At the left, the 0–4–4T 55189 is a former Caledonian Railway shunting pug. In the middle is NBR 0–6–0 65243; on the right a former BR Standard 2–6–4T used on suburban passenger duties. 55189 is better known as CR 419 and is the standard bearer of the Scottish Railway Preservation Society. 65243 is named 'Maude'.

HERITAGE AND PRESERVED STEAM

Britain is uniquely blessed in possessing both a large number of disused railway lines, some running through picturesque parts of the country, and an army of enthusiasts committed to the preservation and care of locomotives and rolling stock. Their efforts mean that admirers of steam locomotives of all ages need not rely only on descriptions in books and their own imaginations to see, feel, hear and smell these magnificent machines. This section tells the remarkable story of preserved steam railways in Britain and provides details of the country's priceless collection of preserved lines and steam railway museums.

HERITAGE AND PRESERVED STEAM

With the inevitable modernization of rail transport from the early 1960s onwards, the railway traveller from those times might well have predicted that the great steam locomotives of the Victorian era and the first half of the twentieth century would soon be nothing, but a memory. At the same time as the curtain was closing on the steam power era, Dr Beeching's 1963 report, *The Reshaping of British Railways*, ensured that the elaborate British railway network– essentially completed by 1880 – would soon be radically reduced. Of course, even if Dr Beeching had never been born, it is inevitable that the thousands of branch lines, which fell to his 'axe', would eventually have been chopped by someone else. Many routes had never returned a dividend to their nineteenth-century shareholders and there were limits to the subsidies which the post-war nationalized rail service could claim. With the growth of mass car ownership and lorry transport, it was not unusual to see virtually empty diesel multiple units (carriages with a built-in engine and a driving cab at both ends) on rural routes.

However, while the all-powerful Victorian railway system has now vanished over the horizon, trailing clouds of steam in its wake, something very tangible remains. In Britain today there is a flourishing preserved locomotive and heritage railway movement. There are three main reasons for this. Firstly, steam locomotives are not easy to dispose of. The hundreds of engines rescued from the Barry Scrapyard and other locations would no doubt have been scrapped eventually, but the time and effort needed gave an opportunity for preservation. Secondly, where old rail routes have not been crossed by new roads or covered by housing estates, the track beds, cuttings and embankments live on and can be brought back into use relatively easily, although the Sunday-afternoon track-laying enthusiast might not agree. Thirdly, and perhaps most significantly, heritage rail and steam has been preserved largely through the efforts of thousands of dedicated volunteers, giving their time, skills and money to preserve a unique part of British history.

PRESERVATION PIONEERS

The survival of very early steam locomotives such as 'Puffing Billy' (built 1813–14), 'Wylam Dilly' (c.1816), 'Locomotion No. 1' (1825) and 'Rocket' (1829) prove that locomotive preservation is hardly a twentieth-century phenomenon.

The preservation history of 'Puffing Billy', the oldest locomotive still in existence in the world, is particularly dramatic. In 1862, it was loaned to the Science Museum by Captain Blackett, of Wylam Colliery, where it had given faithful service since 1814. Unfortunately, a year later, Blackett asked for either the locomotive to be returned, or for a payment of £1,200, which at that time was a great deal of money. This kind of sum was impossible to find and the Curator, Francis Smith, was instructed to dismantle 'Puffing Billy' and return it to Wylam Colliery. Fortunately, Smith made an unauthorized last-minute

offer of £200, which Blackett accepted, securing the Science Museum one of its most popular exhibits. Without Smith's enterprise, 'Puffing Billy' might have ended its life in pieces, rusting away in a corner of the colliery.

Well before the end of the steam era, these early examples of steam locomotives were seen as inherently interesting and put on public display. 'Locomotion No. 1', which hauled the opening-day train on the Stockton and Darlington Railway in 1825, was displayed at William and Alfred Kitching's locomotive workshop at Darlington from 1857 and in 1892, it was exhibited on one of Darlington station's main platforms as an early 'tourist attraction'. A contemporary sketch is kept at the National Railway Museum, York.

GWR prairie tank No. 5199 on the Bluebell Railway (opposite).

Robert Stephenson's 'Invicta' (above).

Stockton and Darlington locomotive 'Derwent', for many years on display at Bank Top Station, Darlington (below).

Other early locomotives were seen as worth preserving, too, such as Robert Stephenson's 'Invicta', built immediately after 'Rocket' for the Canterbury and Whitstable line in Kent, which was the first to use locomotive steam power to convey passengers in regular service. 'Invicta' was exhibited at the Golden Jubilee of the Stockton and Darlington Railway in 1875, then stored for a while, after which it spent around eighty years on display in public gardens in Canterbury. In 1977, it was restored structurally, but not to working order and is now kept at the Canterbury Heritage Centre.

Corfe Castle station in Dorset was originally an intermediate station on the London and South Western Railway branch line from Wareham to Swanage. The line and station were closed by British Rail in 1972. It has since reopened as a station on the Swanage Railway, a heritage railway that currently runs from Norden, just north of Corfe Castle, to Swanage. The painting by Malcolm Root (right) shows ex-LSWR/SR Drummond M7 class 0–4–4T No. 30060 leaving Corfe Castle with a passenger train in the late 1950s.

The photograph (above) shows a Swanage Railway train at Corfe Castle station in November 2006, hauled by BR Standard class 4 locomotive No. 80104.

BRANCH LINES FROM THE PAST

Fiercely competitive Victorian railway companies, determined to build a railway line wherever one could be built, were the motive force behind the growth of the huge nineteenth-century network. Very often, these routes disappeared entirely, until rescued, at least in part, by bands of enthusiasts. Nothing illustrates this process better than the railway history of the Isle of Wight.

The island became a popular place to visit from the 1850s, with tourists following in the steps of Queen Victoria, Prince Albert and Lord Tennyson. The first locomotive-worked route, operated by the Cowes and Newport Railway, opened in 1862 and ran for 7 km (4½ miles) between Newport and Cowes. By 1900, when the network was completed, seven more railway companies had built lines across the Island. In 1966, all steam operations ended, with only an electrified route between Ryde Pier Head and Shanklin remaining in use.

Fortunately, when all the old locomotives and other rolling stock were scrapped, a few items were preserved by the Wight Locomotive Society, which eventually became the Isle of Wight Steam Railway. By 1971, steam trains were running again on a 3 km (2 mile) stretch of line. After many years of effort by volunteers, helped by rail companies and other organizations, the railway was extended to 8 km (5 miles) of the old Ryde and Newport line, which first opened in 1975. The current route passes through the quiet rural island countryside, taking passengers back in time to the heyday of the branch-line railway.

One of the features of heritage railways in Britain is their variety and as well as peaceful countryside routes, there are numerous examples which run through rugged and dramatic industrial landscapes. The Pontypool and Blaenavon Railway is typical of these restored lines. Originally built in 1866 to transport coal to the Midlands, the final stretch from Blaenavon to Pontypool closed in 1980, but started running steam trains again in 1983. The preserved line runs for just over 2 km (1¼ miles) from the terminus, Furnace Sidings station, through industrial remains from the old coalfield to Whistle Halt in the north: there are now plans to extend the line further. The route is the steepest standard gauge preserved passenger line in Britain, and Whistle Halt is the highest station in England and Wales – almost 400 m (1307 ft) above sea level. Steam locomotives currently under restoration include a Great Western Railway 0–6–0 pannier tank locomotive No. 9629, built in 1945. As with so many of these enterprises, the Pontypool and Blaenavon Railway relies on volunteers for its operation.

Hunslet 'Austerity' 0–6–0ST No. WD198 'Royal Engineer' at Havenstreet on the Isle of Wight Steam Railway, May 2009 (left).

Romney station on the Romney, Hythe and Dymchurch narrow gauge railway (opposite).

THE NON-STANDARD GAUGES

Britain's non-standard gauges have very different histories and this is reflected in their preservation. Sadly, it is not possible for the railway enthusiast to make a steam-hauled journey on one of Brunel's Great Western Railway broad gauge routes, as all the lines had been converted to standard gauge by 1892. Re-laying a broad gauge line would be technically possible, but because of the costs involved, this is unlikely ever to happen other than on a short stretch, such as the recreated broad gauge track at the Didcot Railway Centre. Another reason why broad gauge is unlikely to be restored on a large scale, is that unsurprisingly (given the lack of track) only one of the original broad gauge locomotives is still in existence. 'Tiny', a small shunting engine, was built in 1868 and is now preserved at the South Devon Railway Museum. It is unfortunate that we can no longer see two other broad gauge engines, especially since they were at first preserved. 'Lord of the Isles' was an express 4–2–2 locomotive of the 'Iron Duke' class that entered service between 1847 and 1855. 'North Star' was originally built for the New Orleans Railway by Robert Stephenson & Co in 1837. Regauged for the Great Western Railway, it was one of the very first engines that the company possessed, hauling the inaugural train on 31 May 1838. These two locomotives were preserved at Swindon until 1906, when it was decided that they were taking up too much space and they were both scrapped.

The story of narrow gauge is very different, as there was never really any competition with the standard gauge alternative and few were ever converted to the wider track. Narrow gauge was the sensible choice for industrial railways in difficult terrain and it presents some distinct advantages to the preservationist in terms of the costs of the stock and of engineering work. The result is a fine range of heritage narrow gauge throughout Britain.

STEAMING AHEAD

For most enthusiasts, the joy of railway preservation can be summed up in the image of a powerful steam locomotive, resplendent in a long-lost livery: perhaps the deep blue of the Great Eastern Railway, with its elaborate 'lining out', or the distinctive red of the Midland Railway. The large numbers of steam locomotives that have been preserved means that a wide range of representative types is now in working order, on static display or undergoing restoration, from the mighty 4–6–2 Pacifics to the more humble, but no less interesting, tank engines and shunters. Many of these are in big collections, such as that of the National Railway Museum in York, but others can be found all over Britain in smaller museums and heritage railways. Large numbers of the locomotives in the National Collection are 'out on loan' to other museums. One of the earliest types on loan from York is the 'Derwent', a steam locomotive built in 1845 for the Stockton and Darlington Railway by William and Alfred Kitching. It is currently preserved at the Darlington Railway Centre and Museum.

For the person who wants to get even closer to the reality of the steam locomotives of the past, many heritage railways offer a 'footplate experience' – the chance to drive one of these powerful machines. The courses offered by the East Somerset Railway are typical of the kind. Participants receive a safety briefing and an introduction to the mechanics of a steam engine. After assisting in the cab, the 'learner drivers' have an opportunity to take the controls of an empty passenger train, while being closely supervised by a qualified crew. All the additional, but essential, tasks of keeping a steam locomotive in action can be experienced, including 'oiling up' all the moving parts that need lubrication and shovelling coal. A footplate experience brings home to the participant, not just the great skill needed to operate a steam locomotive, but also the very demanding physical nature of the work, very much in the 'heavy industry' tradition of the nineteenth and early twentieth century.

For enthusiasts who favour a longer steam-hauled trip, there are now many excursions and special trains to choose from. The 'Dorset Coast Express', which runs in August and early September, is an example of this flourishing heritage sector. This leaves London Victoria for Weymouth, along the Dorset coast, with a different steam locomotive used for each direction.

The 'Dorset Coast Express' at Upwey Bank (below).

REPLICAS AND REBUILDS

Restored steam locomotives inevitably need a great deal of work before they can be brought back to steam, often involving machining some parts from new. The Barry Scrapyard, the source of many preserved locomotives, allowed preservationists to cannibalize engines from the same class; this system worked well until the 'donor' was itself restored. However, in some cases not even the remnants of the original were available and a decision was made to build the type from scratch. One of the best-known examples of a rebuild is 'Tornado' No. 60163, based on the London and North Eastern Railway 'Peppercorn' A1 class 4–6–2 steam locomotives. 'Tornado' incorporates a good deal of internal modification, mainly to meet safety regulations for regular main-line use, so cannot be considered an exact copy.

The modern electronic safety systems installed on 'Tornado' were, fortunately, not required for the working replica of Stephenson's 'Rocket', which was never intended to haul trains on today's railway network. This replica of the 1829 'Rocket' was built in 1979 for the 150th-anniversary celebrations of the locomotive's first outing. The bright yellow exterior hides some significant differences 'under the bonnet', where modern engineering techniques have been used to prevent any replication of the occasional boiler explosions experienced by locomotives of the 1820s. However, purists should be wary when they point out that this is not the real 'Rocket'. The original model (on display in the Science Museum) had already been extensively modified by 1830 and looks less like the 1829 version than its replica. The replica 'Rocket' has given passenger rides along a 150 m (500 ft) track through Kensington Gardens in Central London and is based at the National Railway Museum in York. Other well-known replicas include working models of 'Locomotion No. 1', built in 1975 to celebrate the 150th anniversary of the opening of the Stockton and Darlington Railway and the broad gauge 'Iron Duke', constructed in 1985.

'Edward Thomas' crossing the Dolgoch Viaduct on the Talyllyn Railway (right).

A LOST WAY OF LIFE

The locomotives of the great era of steam needed far more than a network of track to keep them in business. Steam railways were labour intensive and by 1900, over 600,000 people were employed by the railways. Behind the visible ranks of uniformed station staff, ticket collectors, porters, drivers and footplate crew lay an army of support workers. These ranged from clerks processing vast amounts of paperwork by hand, to skilled engineers, some making new locomotives, others employed in the continual heavy maintenance work needed to keep such large, coal-fired engines running smoothly.

As well as numerous workers, a complex infrastructure was needed to keep steam locomotives under power. Water towers and coal towers, turntables, engine sheds and the web of hand-powered semaphore signals and signal boxes made up an unmistakable 'steam railway landscape'.

Most of the labour force and much of the infrastructure associated with, essentially, Victorian technology has now gone, but thankfully this heritage has been preserved in railway museums and restored railways across Britain. The Steam Museum of the Great Western Railway in Swindon is one of a number of centres, which recreate the atmosphere of the old steam railways, with the sights and sounds of workshops, boiler making, back-room offices, and a station platform.

STEAM PRESERVATION – THE FUTURE

Today, many of the older volunteers and enthusiasts, who keep our preserved steam railway lines in operation, will have their own memories of childhood journeys complete with puffing locomotives, corridor carriages, leather-strap window openers and a choice of 'smoking' or 'no smoking' compartments. However, anyone born after 1970 will have to rely on heritage railways for their experience of steam-hauled trains. Fortunately, preserved steam operations continue to flourish, with plans to extend lines and bring more locomotives back into working order. Established lines such as the West Somerset Railway and the narrow gauge Talyllyn Railway are continuing to extend their routes and develop their infrastructure. There are also new routes: for example, the Aln Valley Railway Society's work to reinstate the 5 km (3 mile) former British Railways branch line from the East Coast main line at Alnmouth to the town of Alnwick in Northumberland. The continued health of preserved steam is important. The nostalgia and romance associated with our historic railway network is a potent force and underlines the cultural significance of the steam train era: one of enterprise, optimism, skill and technological innovation.

Preserved ex-LMS Stanier 'Black 5' No. 44871 running on the East Lancashire Railway.

STANDARD GAUGE HERITAGE RAILWAYS

There are now over a hundred British standard gauge heritage railway lines, which have a track gauge of 1435 mm (4 ft 8½ in), with around fifty offering regular passenger rides between two or more stations. The large majority of these were established by dedicated enthusiasts and rely heavily on unpaid volunteers to keep the lines open. Most run over stretches of lines built before 1900 and which were closed in the early 1960s. Typical examples include the Tanfield Railway (near the Beamish Museum, County Durham) and the Bowes Railway in Gateshead. The Tanfield Railway first opened as a horse-drawn coal wagon line in 1775: its current collection of locomotives includes 'Sir Cecil A. Cochrane' No. 7409, a small 0–4–0ST built by Robert Stephenson and Hawthorns in 1948.

Railway construction took place on a massive scale in Victorian Britain after the opening of the Stockton and Darlington line in 1825. However, by the early 1960s, road transport – for freight as well as passengers – seemed much more modern than rail: worries about fuel economy and carbon emissions lay in the far future. Dr Beeching's 1963 report resulted in the closure of more than two thousand stations and thousands of miles of track.

Demolition of these lines was usually comprehensive, with wholesale removal of rails, signalling gear and bridges. Some old trackbeds remain as popular walks, such as the route of the former Princetown branch line across Dartmoor; others, such as the Glastonbury to Burnham-on-Sea line in Somerset, are now increasingly hard to identify.

Fortunately, many enthusiasts have recognized the importance of these 'redundant' lines as part of Britain's cultural heritage and have successfully preserved numerous examples for future generations to enjoy.

BLUEBELL RAILWAY *East Sussex*

The Lewes and East Grinstead Railway opened in 1882, carrying a wide variety of passengers and goods. It finally closed in March 1958. An ambitious plan to reopen the whole line with a diesel service proved unworkable. It was followed by a successful scheme to reopen a shorter stretch of the route, managed by the Bluebell Railway Preservation Society. The first train ran in August 1960, hauled by a Stroudley 'Terrier' 0–6–0T locomotive, bought from British Railways and originally built in 1875. A regular passenger service has operated ever since. The Bluebell Railway now runs for nine miles on the border between East and West Sussex.

The Lewes and East Grinstead line was first closed in 1955, but a famous legal challenge was made by a local resident, Miss Bessemer. She discovered an 1878 Act of Parliament, which said that the line had to be kept open – and British Railways were forced to reopen the line in 1956. After a Public Inquiry, Parliament was forced to repeal the Act in order to defeat her.

The three stations on the Bluebell Line are Kingscote, Horsted Keynes and Sheffield Park. Kingscote is a quiet country station which has been largely restored to its British Railways 1950s condition and is accessible by bus from East Grinstead – there is no public car parking. Horsted Keyes is a large country station, housing the carriage and wagon department, where rolling stock is restored and maintained. Sheffield Park is the home of the locomotive department.

All passenger trains are hauled by steam locomotives. The Bluebell Railway has a large collection of heritage locomotives, carriages, vans and goods wagons, alongside refurbished signal boxes, signals and historic railway buildings.

EAST LANCASHIRE RAILWAY

When the East Lancashire Railway opened in 1846, it was a busy passenger and freight route from Bury to Rawtenstall and beyond. The Bury–Rawtenstall line continued to carry passengers until 1972, when British Rail withdrew the service. Complete closure took place in 1980, after the remaining freight services were discontinued.

The first section of the preserved line was reopened in 1987, when 6 km (4 miles) of track provided regular passenger services between Bury and Ramsbottom. Following further expansion, the line is now over 19 km (12 miles) long and runs between Heywood and Rawtenstall in East Lancashire, passing over viaducts and through tunnels. It is managed by the East Lancashire Railway Preservation Society (originally formed in 1968) in partnership with the East Lancashire Light Railway Company and the East Lancashire

Railway Trust. In addition to the stations at Heywood and Rawtenstall, there are intermediate stations at Bury (Bolton Street) and Ramsbottom, and two unstaffed halts at Summerseat and Rawtenstall. Bolton Street station was rebuilt in the 1880s and the existing platform canopy dates from that time. Other historic features on the line include a level crossing at Ramsbottom station, which retains its wooden gates worked by the traditional 'ship's wheel' in the adjacent signal box.

There are regular passenger services, which are mainly steam hauled, although diesel, diesel electric, diesel hydraulic and shunting locomotives are also preserved. Many of the steam locomotives based at the East Lancashire Railway are main-line certified, including the 'Jubilee' class No. 5690 'Leander', built at Crewe in 1936.

GLOUCESTERSHIRE WARWICKSHIRE RAILWAY

The current Gloucestershire Warwickshire Railway, also known as 'The Honeybourne Line', runs along part of the old Great Western Railway's main line from Birmingham to Cheltenham. Built mainly between 1900 and 1906, the line carried passengers and freight, including farm produce from the Vale of Evesham. It closed to passenger traffic in 1960 and the last goods train ran in 1976.

By the time the Gloucestershire Warwickshire Railway Preservation Group (formerly the Gloucestershire and Warwickshire Steam Railway) had taken over the line in 1981, all the rails and many of the buildings had been removed. The first passenger train ran on the restored line in 1984, although only on 400 metres (quarter of a mile) of track. Since then, volunteers have gradually restored the line, replacing track, station buildings, signal boxes and signals and now 16 km (10 miles) of track run from Toddington to Cheltenham Race Course. There are regular passenger services, many of which are hauled by steam locomotives.

The Gloucestershire Warwickshire Railway runs through Cotswold countryside and the headquarters of the railway, Toddington Station, houses the locomotive sheds. There are three other stations on the line: Winchcombe (which is located in the village of Greet); Gotherington Halt, which can only be accessed by rail, or on foot; and Cheltenham Race Course.

One of the operational locomotives on the line is No. 7903 'Foremarke Hall', built in 1949. In 1951, this became the first locomotive to travel from London to Plymouth in less than four hours. Carriages include a refurbished restaurant/buffet car, No. 1675, built in 1959.

GREAT CENTRAL RAILWAY *Leicestershire*

The 'London Extension', as the Great Central Railway was then known, opened in 1899 as an extension of the Manchester, Sheffield and Lincolnshire Railway and was a busy main-line service linking Sheffield to London. The ambitious chairman, Edward Watkin, wanted to extend the line further to Europe via a Channel tunnel and managed to start some excavations before the money ran out. Long sections were closed in 1966 and the final Nottingham–Rugby stretch was discontinued in 1969.

A preservation society was formed in the year of closure and gradually more track was brought into use. The preserved section of 13 km (8 miles), running between Loughborough and Leicester, is now fully open. As a double-track, main-line heritage railway, the current Great Central Railway is one of the few places left in the world where full-sized steam engines can be seen passing each other at speed. Many of the regular passenger services are steam hauled.

There are four stations on the line, which runs through Leicestershire. Loughborough Central is the main station for the route. Typical of the 1960s, it is home to a museum and the engine shed. Quorn and Woodhouse station has a 1940s style, and Rothley has some Edwardian features. Leicester North is the southernmost station, with limited parking.

The Great Central Railway has a large range of home-based locomotives, including a heavy freight 'Robinson' class 04 2–8–0 No. 63601. There are also regular 'visiting' locomotives and a wide selection of carriages, including dining cars.

KEIGHLEY AND WORTH VALLEY RAILWAY *West Yorkshire*

The original line, built to service the flourishing textile industry of the time, opened in 1867. Trains hauled hundreds of tons of coal up the Worth Valley each week for the steam-powered looms. The line was closed by British Railways in 1962 and a preservation society was soon formed, with the line reopening to passenger traffic in 1968. The society decided, as far as possible, to recreate the atmosphere of a 1950s branch line.

Located in West Yorkshire, the railway comprises an 8 km (5 mile) stretch of track from Keighley to Oxenhope, including 6 stations. Keighley Station is owned by Network Rail, who lease two platforms to the Keighley and Worth Valley Railway: one of them has the original waiting room and toilets, which are still in use. Oakworth Station, which has retained most of its original features, was used in the 1970 film version of Edith Nesbit's book, *The Railway Children*. This resulted in such a busy season that extra trains had to be laid on for the tourists in 1971. Howarth Station is the main base for locomotives and is linked to the village by a footbridge. There are three further stations at Ingrow (West), Damems (which claims to be Britain's smallest station) and Oxenthorpe. Many of the regular passenger services are steam hauled.

The railway is home to a wide variety of locomotives, carriages and railbuses. Locomotives in use on the line include a Taff Vale Railway 0-6-2 No. 85, built in 1899 to haul coal in South Wales and restored in 2000.

KENT AND EAST SUSSEX RAILWAY

The first section of the Kent and East Sussex Railway (initially known as the Rother Valley Railway) opened in 1900 and the original line was completed in 1905. Unusually, it was not absorbed into one of the four great companies created in 1923 and remained independent until all the railways were nationalized in 1948. The line was closed to passengers in 1954, apart from a few trains for hop-pickers in the summer, although goods services continued until 1961, when the railway was closed.

A preservation society was formed soon after closure and after years of negotiations with the Ministry of Transport, the first 3 km (2 miles) of track opened in 1974, with a final extension in 2000.

The current line runs for just over 16 km (10 miles) through the Rother Valley, on the border of East Sussex and Kent. There are regular passenger services, most hauled by steam locomotives, with both vintage and British Railways coaches. The original line, although standard gauge, was originally built to 'light railway' specifications, which encouraged cheap construction of lines in remote rural areas. This has meant that many features, such as embankments, have been updated since preservation.

There are five stations on the route. Tenterden Town is the main centre for the preserved railway, with a museum and carriage workshop. Rolvenden Station houses the engineering centre with locomotive and carriage sheds. Wittersham Road features an 1882 signal box (originally located in Dover). There is a range of original station buildings at Northiam and at the terminus, Bodium.

The railway is home to a wide variety of steam and diesel locomotives, coaching stock and wagons. Locomotives include an 0–6–0T 'Terrier' class, which first entered service in 1875. When sold in 1901 for £650 it had already travelled over 1,000,000 km (620,000 miles).

NENE VALLEY RAILWAY *Cambridgeshire*

The original line on this route was constructed by the London and Birmingham Railway in 1845 and the Nene Valley Railway of today comprises the eastern part of this line. After passing through several ownerships, the line was partially closed after 1964, with complete closure in 1972.

In 1974, the Peterborough Development Corporation bought a stretch of the line and leased it to the Peterborough Railway Society, who operated the heritage Nene Valley Railway. The first train ran in 1977, after the line was upgraded.

The current route covers 12 km (7.5 miles), terminating at Peterborough, in Cambridgeshire. There are five stations on the route. Wansford is the headquarters of the Nene Valley Railway and historic features include a 1907 signal box, originally equipped

with sixty levers – one of the largest preserved signal boxes on its original site. Peterborough (Nene Valley) station is the centre for goods wagon restoration and houses an independently-run museum. The three other stations are Yarwell Junction, Ferry Meadows and Orton Mere.

When the Nene Valley Railway was first established as a heritage project, it was decided that the line would be upgraded to a Continental standard 'loading gauge' – that is, to accommodate the additional width and height of Continental trains. As a result, it is now home to a very varied collection of locomotives and rolling stock, including examples from Britain, France, Germany, Italy, Denmark, Sweden and Norway. British locomotives include the BR 5MT 4–6–0 No. 73050 'City of Peterborough', built in 1954, which first saw service on the Somerset and Dorset line between Bath and Bournemouth.

NORTH YORKSHIRE MOORS RAILWAY

The North Yorkshire Moors Railway originally opened in 1836 as a horse-worked track, the Whitby to Pickering Railway, after a report and recommendations by George Stephenson. The first carriages were essentially stage coaches adapted for railway lines, but able to carry passengers much more efficiently than on the roads of the time. After absorption by the York and North Midland Railway, it was rebuilt for steam traction from 1845. Following several changes of ownership, it eventually became victim to the 1963 Beeching Report and British Railways closed the line to passenger services in 1965, with some goods haulage surviving for a further year.

A preservation society was first formed in 1967, carrying out some maintenance work and running occasional 'Steam Galas'. The current North Yorkshire Moors Railway opened in 1973. The 29 km (18 mile) line runs through the scenic North York Moors National Park, between Pickering, a market town,

and the village of Grosmont. There are five stations on the line. Pickering has been restored to its 1937 condition, with a range of original fittings. A walkers' request stop operates at Newton Dale Halt and Levisham is also popular with walkers. Goathland has a building with authentic furniture and artifacts and a restored coal loading facility. Grosmont station is the terminus for the North Yorkshire Moors Railway, housing the locomotive sheds. The station building has been restored to the British Railways' style of the 1960s.

There are regular passenger services, many hauled by steam locomotives, with a wide range of heritage rolling stock. Operational locomotives include the LNWR class G2 (Super D) 0–8–0 No. 49395, the 'super' standing for superheated.

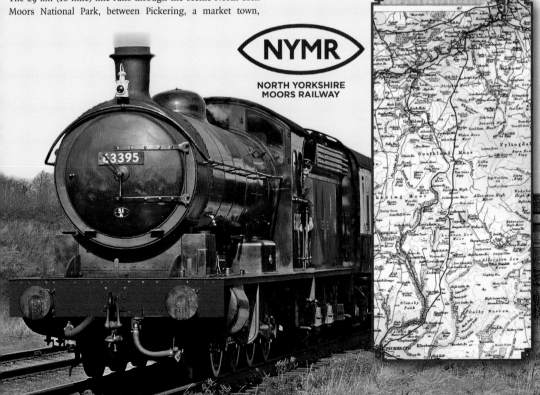

SEVERN VALLEY RAILWAY

The original Severn Valley Railway opened in 1862 and was absorbed into the Great Western Railway in the 1870s. The line was closed to through passenger and freight operation in 1963, with full closure in 1970, when a few residual services were discontinued.

Preservation of the Severn Valley Railway started in 1965, when the Severn Valley Railway Society was formed at Kidderminster. The first section of line was opened for public passenger services in 1970 and finally extended to Kidderminster in 1984.

The current railway runs between Kidderminster, in Worcestershire and Bridgnorth, in Shropshire, covering 26 km (16 miles), much of it following the course of the River Severn. The river is crossed at one point by the 60 m (200 ft) single-span Victoria Bridge. There are regular steam-hauled passenger trains, connecting the five stations on the line, plus a halt. Bridgenorth station has a view of the locomotive works from the footbridge. Hampton Loade is a country station with original features. Highley is a carefully-restored stone-built station, again set in countryside, which also houses the engine shed. Arley has a restored signal box and station gardens. Bewdley is a large former country-junction station, with buildings to match. Kidderminster has some original station buildings and there is a railway museum nearby.

There is a boiler repair shop at Bridgenorth and carriage maintenance and repair facilities at Bewdley and Kidderminster. This allows major locomotive and rolling stock restoration to take place. The preserved steam locomotives used on the route include a GWR/BR 0–6–0 Pannier Tank No. 7714, a type built between 1929 and 1950.

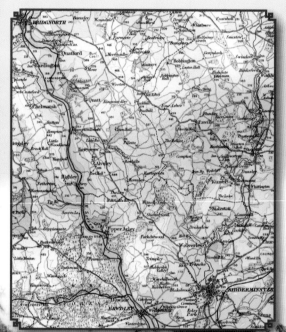

SOUTH DEVON RAILWAY

The current South Devon Railway runs on part of what was once the Buckfastleigh, Totnes and South Devon Railway, which used Brunel's broad gauge track and opened in 1872. The line was taken over by the Great Western Railway in 1876 and converted to the narrower standard gauge in 1892 – over one weekend. The last passenger train ran in 1958, with freight continuing until 1962.

The line was reopened by a commercial company, the Dart Valley Light Railway Ltd, as a steam-operated branch line. The first passenger trains ran in 1969, with Dr Beeching conducting the opening ceremony. In 1991, the line was taken over by the South Devon Railway Trust and renamed the South Devon Railway. The heritage railway now runs regular steam passenger trains for 11 km (7 miles) along the valley of the River Dart, between Buckfastleigh and Totnes, including three stations. Totnes (Littlehempston) station, on the east bank of the River Dart, is close to the Totnes main-line station and is in rural surroundings, although close to the centre of the town

of Totnes. Buckfastleigh station became the terminus of the line in 1971, when the route beyond the station was closed for improvements to the A38 road. There is also an intermediate station at Staverton.

The railway maintains a fleet of historic rolling stock, including steam and diesel locomotives and a well-equipped engineering workshop. Locomotives on exhibition include 'Tiny', a peculiar (to modern eyes) broad gauge locomotive, built in 1868. The railway is also home to GWR 4–6–0 No. 4929 'Dumbleton Hall', built in 1929.

WATERCRESS LINE (MID-HANTS RAILWAY)

The Mid-Hants Railway is known popularly as the 'Watercress Line', from the days when it was used to transport local watercress to London. The Mid-Hants Railway opened in 1865, providing an alternative route between London and Southampton. With nationalization, the line became part of the Southern Region of British Railways in 1948. Although it survived the Beeching axe, in 1973 it was finally closed by British Railways.

A group of enthusiasts raised enough money to reopen part of the line in 1977, and the final section of the current track began operations by 1985. The route is now managed by the Mid-Hants Railway Preservation Society. There are regular steam-hauled passenger services, supplemented by vintage bus rides.

The line runs for 16 km (10 miles) though Hampshire and includes four stations. Alton is at the northern end of the line and has a cross-platform connection to London Waterloo. Meadstead and Four Marks, the highest station in Hampshire, houses the signal and telegraph department. It has been restored to resemble a rural 1940s railway station.

Ropley houses the locomotive sheds, and is the centre for engineering on the line. The yew topiary in the station gardens has been in place for more than a hundred years. Aresford station, at the southern end of the line, is the main passenger terminus and is also home to a museum.

There is a variety of heritage rolling stock and steam locomotives based at the Watercress Line, including the powerful SR 4–6–0 No. 850 'Lord Nelson', built in 1929 as the first engine of the 'Lord Nelson' class (all named after famous admirals).

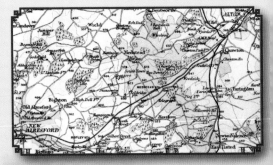

WEST SOMERSET RAILWAY

The West Somerset Railway, built to Brunel's broad gauge, first opened in 1862 and by 1874, it linked Taunton with Minehead. The line benefited from tourist traffic to the Somerset coast and the opening of a Butlins holiday camp at Minehead in 1962, but this did not prevent eventual closure in 1971. Supported by the Minehead Railway Preservation Society, the line was reopened as a heritage railway. The first section of line reopened in 1976 and the current route, including a number of new stations, was operational in 2009.

The 36 km (22 mile) line between Taunton and Minehead is the longest heritage rail line in the country, although regular passenger services only operate on the 31 km (19 mile) stretch between Bishops Lydeard and Minehead. There are ten stations on this stretch of the line, including Bishops Lydeard, where there is a visitor centre, museum and locomotive servicing depot and Crowcombe Heathfield, where the station is designated as a Grade II listed building. A signal box dating from 1875 can be found at Williton station and the stone station building at Dunster is also Grade II listed. Dunster station provides a workshop for the civil engineering team on the line. Minehead station is the headquarters of the West Somerset Railway and houses the main departments, including engineering and signals.

The line operates a wide range of steam and diesel locomotives and heritage rolling stock. These include a GWR 280 class 2–8–0 No. 2874, one of a class of heavy-freight locomotives built between 1903 and 1919.

NARROW GAUGE HERITAGE RAILWAYS

For many people, the term 'heritage railway' will conjure up a picture of a standard gauge train headed by a full-sized steam locomotive. However, there are over forty narrow gauge heritage lines operating in Britain today that offer passenger services.

The history of narrow gauge railways is as interesting as that of the more common standard gauge, but with some significant differences. Most were built for industrial use, although others, such as the Ffestiniog Railway in North Wales, also carried passengers. Large building projects often involved the short-term use of a railway of this kind. A good example is the Nidd Valley Light Railway, built to service the construction of Scar House Reservoir in North Yorkshire and subsequently dismantled. However, not all narrow gauge railways have links with Britain's industrial heritage. Some, such as the Snowdon Mountain Railway and the Romney, Hythe and Dymchurch Railway, were opened specifically for passengers: the latter is well known for its one-third scale 'Pacific' locomotives.

The advantages of using a narrow gauge – usually defined as less than 1067 mm (3 ft 6 in) – are cheapness and ease of construction. The disadvantages of lower speeds and reduced carrying capacity were less significant on short industrial routes. Unsurprisingly, many of these railways were built in the coal and slate mining areas of Wales, which now has a fine selection of narrow gauge preserved lines.

The preservation of these railways has followed a familiar pattern. After originally being built for horse-drawn wagons, they were then converted to steam traction in the mid-1800s. Many were later abandoned when the quarries ceased working, or when road haulage became feasible – but fortunately, some excellent examples have been rescued for posterity by committed enthusiasts.

A train on the Welsh Highland Railway climbs past Llyn Cwellyn towards Rhyd Ddu station, reopened in August 2003.

FFESTINIOG RAILWAY *North Wales*

Founded in 1832, the Ffestiniog Railway is the oldest independent railway company in the world. Built to transport slate from the quarries in the Welsh mountains around Blaenau Ffestiniog, the railway used gravity to take the wagons down to the coast and horses to pull them back up again. Steam locomotives were introduced in 1864. By the 1920s, the railway depended more on tourism than it did on the slate industry and the line finally closed in 1946.

In 1954, a group of rail enthusiasts bought the company, which eventually became the Ffestiniog Railway Trust. Shortly afterwards, restoration of the line was hampered by the creation of the Tanygrisiau reservoir (Llyn Ystradau) for an electricity generation scheme, which flooded part of the northern stretch of track – compensation for this was finally awarded eighteen years later. The first public passenger trains ran on part of the route in 1955 and, after extensive phases of reconstruction and restoration, including a deviation a round the new lake, the current railway was completed in 1982.

Regular steam-hauled passenger services operate on the Ffestiniog Railway, which now runs for 22 km (13½ miles) from Porthmadog harbour to the town of Blaenau Ffestiniog. The five main stations include Porthmadog, where the original building dates back to 1879, and Minffordd station, which houses a carriage shed. Further down the line, Penrhyn station has been largely restored to its nineteenth-century condition. Tan-y-Bwlch is set in the Snowdonia National Park and the line terminates at the new Blaenau Ffestiniog station, which opened in 1982.

The route runs, as it has always done, on narrow gauge track of 600 mm (just under 2 ft). There is a large selection of narrow gaugeheritage locomotives and rolling stock, with a number of 'Double Fairlie' examples, including 'Merddin Emrys', built in 1879.

LEIGHTON BUZZARD RAILWAY *Bedfordshire*

The original line, the Leighton Buzzard Light Railway Ltd in Bedfordshire, was established in 1919 to transport the top-grade silicate sand from local quarries to the nearby LNWR Dunstable branch line and the Grand Junction Canal. It was built mostly from surplus materials from the 'War Department Light Railways', which had provided rail connections to the battle zones in France. In the early 1950s, twenty train loads of sand were carried each day, but by 1969, through trains to Dunstable were discontinued and all internal operations ceased in 1981.

In 1967, a group of railway enthusiasts ran occasional passenger trips over the line and regular steam-powered services began in 1968, managed by the Leighton Buzzard Narrow Gauge Railway Society.

The current heritage railway is 5 km (3 miles) long and runs regular steam-powered passenger services from Pages Park, in the southern area of Leighton Buzzard, to Stonehenge Works

in the north – with the usual option of a round trip. The route passes through a mixture of urban areas and countryside. Pages Park houses the locomotive shed, linked to the platform by a footpath from where locomotives can be viewed. Stonehenge Works is the engineering workshop of the preserved railway.

The Leighton Buzzard Railway's locomotive fleet totals over fifty and is one of the largest narrow gauge collections in the United Kingdom. The stock list includes 'Chaloner', a 0–4–0 VBT (vertical-boiler tank) locomotive built in 1877.

Except on its opening in 1919, when a few VIPs were accommodated in special modified wagons, the original Leighton Buzzard Light Railway never carried passengers. This has meant that the heritage railway has had to obtain, or build some unusual conversions. For example, one carriage started life in 1958 as a standard gauge diesel railcar, before being converted to an unpowered, narrow gauge bogie coach.

ROMNEY, HYTHE AND DYMCHURCH RAILWAY *Kent*

The Romney, Hythe and Dymchurch Railway opened in 1927, running on what was then the narrowest gauge in existence for a public passenger service – 381 mm (15 in). It was planned and funded by two millionaires, both racing drivers and railway enthusiasts: Captain J. E. P. Howey and Count Louis Zborowski. Sadly, Count Zborowski was killed in a racing accident before the railway became operational. The line soon became well known as 'The Smallest Public Railway in the World'. It was taken over by the military in World War II and the first section reopened in 1946. As well as catering for enthusiasts, the current railway operates a public transport service, including taking children to and from school. Much needed updating was undertaken in the 1970s, including renewal of bridges and rolling stock. The current railway is supported by the Romney, Hythe and Dymchurch Railway Association.

The 22 km (13½ mile) line runs from Hythe to Dungeness, in Kent, with seven stations along the route. These include Hythe station, where there is a shunting line and turntable, used by locomotives to turn round for the return journey. Dymchurch station is home to the original gents' toilets, designed in 1924. New Romney station is the main base for the railway and was the first station to be built: during construction in 1926, it was inspected by HRH Duke of York. The station houses the locomotive and carriage sheds, and the workshops. Dungeness, at the southern end of the line, is situated in a protected area of shingle headland, with rare plants and wildlife.

The railway operates a fleet of steam and diesel locomotives, built to one-third full size. All the steam locomotives were built between 1925 and 1927. These include 'Hercules', a 4–8–2 'Mountain' class, built by David Paxman & Co in 1927.

THE TALYLLYN RAILWAY *North Wales*

The Talyllyn Railway opened in 1865, to carry slate from the local quarries to the Welsh coast using steam locomotives. Unusually, it was also authorized to carry passengers. Slate traffic finished in 1946, but some passenger services continued to run, despite the fact that the Talyllyn Railway had not been absorbed by British Railways after nationalization. When closure seemed likely in 1951, the line was taken over by the newly-formed Talyllyn Railway Preservation Society, the first ever organization of its kind, founded and driven by the famous engineer and author L. T. C. (Tom) Rolt.

The current railway runs for 12 km (7½ miles) through the Fathew Valley in Gwynedd, Wales, from Tywyn on the west coast to just beyond Abergynolwyn in the northeast. There are regular steam-hauled passenger services to the seven main stations on the route and a number of halts. Tywyn Wharf on the west coast is the railway's main station, where most passengers join the train and is close to the main-line station in Tywyn. Pendre, a small station close to Tywyn, is the operational centre of the railway, home to the locomotive and carriage sheds. Rhydyronen and Brynglas stations are good starting points for walks and Dolgoch Falls is close to a series of waterfalls. Abergynolwyn is the main inland station.

The steam locomotives used on the line include the two that were operated when it first opened, although both have been extensively overhauled over the years. No. 1 'Talyllyn' was the first locomotive in use, built in 1864 as an 0–4–0ST, later converted to an 0–4–2ST (saddle tank). No. 2 'Dolgoch', built in 1866, was another 0–4–0 tank engine, this time with a well tank between the frames and a back tank behind the cab.

VALE OF RHEIDOL RAILWAY *Mid Wales*

The Vale of Rheidol Railway opened in to the general public in 1902, designed to carry timber and ore from the Rheidol Valley to the Welsh coast, although passengers, especially tourists, soon became the main traffic. After eventually becoming part of the Great Western Railway in 1923, the line was taken over by British Railways in 1948 and it became the last steam railway owned by British Rail, until it was privatized in 1989. The railway is now owned by a charity, the Phyllis Rampton Narrow Gauge Railway Trust.

The Vale of Rheidol Railway runs for 19 km (12 miles) between Aberystwyth and Devil's Bridge, in Ceredigion, Wales. There are five main stations and a number of halts. At the western coastal terminus, Aberystwyth, the line shares part of the British Rail main-line station. Capel Bangor station has sidings which link the line to a train shed and Aberffrwd has a water tower for the steam locomotives to top up. Devil's Bridge station, the eastern terminus, has two low-level platforms, a water tower and a number of station buildings.

The track gauge is narrower than is usual for such railways: only 603 mm (just under 2 ft). This gauge was chosen because of the difficulties in construction through the mountainous terrain and there are many sharp curves and steep gradients on the line. Current steam locomotives hauling the regular passenger services include 'Prince of Wales', a 2–6–2 tank engine No. 9. It was newly built for the GWR shortly after their take-over of the railway in 1923, although for some years it was thought to be one of the original 2–6–2T locomotives (No. 1213), which was used when the line first operated.

WELSH HIGHLAND RAILWAY *North Wales*

The current heritage Welsh Highland Railway has its origins in a number of nineteenth-century narrow gauge railways. These included the Nantlle horse-drawn railway, which opened in 1828; the Croesor horse-drawn tramway, established in 1864; and the North Wales Narrow Gauge Railways, constructed between 1873 and 1881. The route was managed by the original Welsh Highland Railway from 1922 and finally closed in 1937.

In 1961, a group of enthusiasts formed a preservation society, which became the Welsh Highland Light Railway. A base was eventually established on a former standard gauge siding in Porthmadog, and after a short length of line was laid, a public service commenced in 1980. By 2009, the route of the Welsh Highland Railway was 31 km (19½ miles) long, from Caernarfon to Beddgelert, through Snowdonia. A recent extension towards Porthmadog and linking with the Ffestiniog Railway, makes a total run of 42 km (26 miles).

There are six main stations on the line, plus a number of halts. Caernarfon is the northern terminus, opening in 1997. Dinas station includes two buildings from the original North Wales Narrow Gauge Railway – the goods shed and the original station building – and is home to the locomotive depot, carriage sheds and railway offices. Waunfawr station opened in 2000. Rhyd Ddu station, which dates from 1881, closed in 1936 and reopened in 2003. The rebuilt Beddgelert station, until recently the southern terminus, connects to Porthmadog station with the newly extended line.

There are regular steam-hauled services, with a range of vintage rolling stock, mainly using a set of four Beyer-Garratt NGG16 (2–6–2+2–6–2) locomotives obtained from South Africa – probably the most powerful narrow gauge steam locomotive type in the world. The power is needed to cope with the difficult gradients along the route.

WELSHPOOL AND LLANFAIR LIGHT RAILWAY Mid Wales

The Welshpool and Llanfair Light Railway opened in 1903 to link the market town of Welshpool to the village of Llanfair Caereinion, a rural community in Powys, Wales. It was originally operated by the Cambrian Railways and became part of the Great Western Railway in 1923 until nationalization. It never made a profit. In 1931, the line lost its passenger services and was finally closed to freight by British Railways in 1956.

Part of the western sector of the line was restored and reopened by a group of enthusiasts in 1963 and the current route was completed in 1981. The line now runs for 14 km (almost 9 miles) westward from a new terminus at Welshpool (Raven Square). The original terminus at Welshpool was alongside the main line station and there are plans for that route to be reinstated at some future date. The other stations on the line (apart from a number of halts) are at Castle Caereinion and at the westernmost terminus, Llanfair Caereinion, where the railway follows the River Banwy on its final stretch.

All passenger services are steam hauled. Locomotives include two of the original engines used on the line, 'The Earl' and 'The Countess', both 0–6–0T types, built by Beyer Peacock in 1902. To supplement the original stock, locomotives have been obtained from a number of foreign sources, including Sierra Leone and Finland. The carriages also include imports: some in regular use, obtained from Hungary and Austria, have an enclosed seating area and an open balcony.

Double Fairlie locomotive 'Merddin Emrys' at Porthmadog on the Ffestiniog Railway (opposite).

PRESERVED STEAM LOCOMOTIVES

A fascination for steam locomotives, and a desire to save some of them for future generations to appreciate, has led to many fine examples being rescued from destruction. Preservation was given an impetus when, in 1955, British Railways made the decision to get rid of steam power altogether, a process that was well underway by the early 1960s.

The journey from working locomotive to successful preservation has often been complex, as in the case of 'Princess Margaret Rose' No. 6203, a London, Midland and Scottish 'Princess Royal class' 4–6–2, built in 1935 and withdrawn in 1962. After withdrawal it was bought by Billy Butlin for static display at the Pwllheli holiday camp in North Wales – via some superficial 'restoration' at Crewe. In 1975, it moved to the Midland Railway Centre in Derbyshire, but a return to working order was not achieved until 1990.

Many other such projects have required years of tenacity and dedication. The preservation of steam locomotive No. 2807, a Churchwood 2–8–0 heavy freight engine, built by the Great Western Railway at Swindon in 1905, is another example. No. 2807 was withdrawn in 1963 and remained in the Barry Scrapyard until 1981, when it was rescued by its current owners, a group who operate as Cotswold Steam Preservation Limited (CSP). Restoration of No. 2807 has been a very long and complex process, but the CSP is confident that this locomotive will soon be steaming again. Similar dedicated enthusiasts have helped to ensure the survival of many other historic locomotives.

Stanier Pacific No. 6233 'Duchess of Sutherland' steams past Monk Fryston, Yorkshire, with an enthusiasts' special, April 2008.

'CAERPHILLY CASTLE'

'Caerphilly Castle' No. 4073 entered service at Paddington station on 23 August 1923. It was the first of the 4–6–0 'Castle' class locomotives designed by Charles Collett, then Chief Mechanical Engineer for the Great Western Railway. Withdrawn from service in 1960, it was first kept in the Science Museum: the transportation by road through central London of the great locomotive generated much interest. 'Caerphilly Castle' is now preserved at the Swindon Steam Railway Museum.

During 1924, 'Caerphilly Castle' was one of the main attractions at The British Empire Exhibition, Wembley, together with the 'Flying Scotsman'. During the subsequent 'Locomotive Interchange Trials' between the Great Western Railway and the London and North Eastern Railway, the 'Castle' class proved superior to all the competing engines.

Collett's design was very successful and 171 'Castle' class locomotives were built between 1923 and 1950. The design was based on the famous 'Star' class of 1907, with a bigger boiler and enlarged cylinders to haul the increasingly heavy trains. As well as haulage capacity, the Castles were noted for their sustained high-speed running and remarkable fuel and water efficiency.

The Castle design was further improved after 1946, with double chimneys fitted to some locomotives and large superheaters for increased speed. In 1958, the 'The Bristolian' express, hauled by 'Drysllyan Castle' No. 7018, reached 160 km (100 miles) per hour. 'Caerphilly Castle' and the others in the class were used to haul passenger express trains, first for the Great Western Railway and then later for British Railways Western Region. They remained on top-line express duties, until they were replaced by diesels in the early 1960s.

'CITY OF TRURO'

'City of Truro' No. 3440 was built in 1903, one of the 4–4–0 'City' class locomotives designed by George Churchward for the Great Western Railway. The working life of this class was comparatively short and, when it was withdrawn from service in 1931, 'City of Truro' was one of the last two remaining. 'City of Truro' is preserved by the National Railway Museum in York.

'City of Truro' is best known for the speed record that it set on 9 May 1904, when it hauled the 'Ocean Mail Express' between Plymouth and Paddington. The service normally carried no passengers, but on this occasion Charles Rous-Marten, a railway commentator and journalist, had been invited on the journey, probably because a high-speed run had been planned. By using milepost timings, Rous-Marten recorded a top speed of 164 km (102 miles) per hour down Wellington Bank in Somerset. The speed was not officially confirmed by the Great Western Railway until 1922 – perhaps because of safety concerns – but after that the record was given much publicity.

Because of its historic significance, when it ceased operations in 1931 'City of Truro' was displayed at the new museum in York. It returned to service in 1957 for British Railways Western Region, and was based at Didcot, where it was used for excursion trains, as well as normal services. After being withdrawn for a second time in 1961, it was kept at the Swindon Steam Museum until returning to York in 1984. For some years, the National Railway Museum has allowed 'City of Truro' to be based semi-permanently at the heritage Gloucestershire Warwickshire Railway.

'DUCHESS OF SUTHERLAND'

'Duchess of Sutherland' No. 6233 is a 'Princess Coronation' class 4–6–2 four-cylinder Pacific, built in 1938 at Crewe by the London, Midland and Scottish Railway and designed by William (later Sir William) Stanier. It was withdrawn from traffic in 1963.

This powerful design represented the height of four-cylinder locomotive design in Britain and was developed from the earlier 'Princess Royal' class. The boiler and cylinders were improved and a 40-element superheater added. The 'Princess Coronation' class could haul large loads at speed and often reached 160 km (100 miles) per hour in normal service. Many were first built as streamlined locomotives, covered with a bathtub-shaped outer skin, but the 'Duchess of Sutherland' was one of the non-streamlined models.

For the first six years of service, 'Duchess of Sutherland' worked trains out of Euston on the West Coast main line. After transfer to the Crewe depot in 1944, it spent its last days in service at Liverpool Edge Hill. Sir Billy Butlin bought the locomotive in 1964 for static display at a holiday camp in Ayr. Following a move to the Bressingham Railway Museum in Norfolk, 'Duchess of Sutherland' was bought by the Princess Royal Class Locomotive Trust in 1996 and was subsequently restored to full working order. On 11 June 2002, it hauled the Royal Train (the first steam locomotive to do so for 35 years) for the Golden Jubilee Tour of Queen Elizabeth II and the Duke of Edinburgh and in 2005, 'Duchess of Sutherland' transported Prince Charles on the Settle–Carlisle Railway – including a spell with the prince at the controls.

'DUKE OF GLOUCESTER'

The Standard class 8P 4–6–2 Pacific 'Duke of Gloucester' No. 71000 was completed at British Railways' Crewe works in 1954 and is usually referred to as the 'Duke'. After only eight years in service, the 'Duke' was withdrawn in 1962. Only one locomotive of this type was ever built. Once the valve gear had been removed for display in the Science Museum, the 'Duke' was sent to the Barry Scrapyard. The Duke of Gloucester Steam Locomotive Trust eventually acquired the decayed remains in 1973 and following restoration and improvements, the 'Duke' was returned to running condition in 1986.

The 'Duke' has a unique and scientifically-important history. The locomotive engineer, Robert Riddles, had wanted to develop a new locomotive in the same power class as 'Princess Anne' No. 46202, but was turned down on grounds of cost. It was only when the

'Princess Anne' was destroyed in a terrible accident in 1952, creating a gap in the roster, that Riddles received the go-ahead for his revolutionary design. New features were incorporated, including an improved British version of valve gear, first designed by Arturo Caprotti in Italy in 1911, combined with three cylinders and an advanced exhaust system. Unfortunately, errors in construction (some of which only came to light during the later restoration) meant that the 'Duke' never performed well, leading to an early withdrawal from service.

During the thirteen years of restoration, the faults in the original locomotive were identified and corrected. When the 'Duke' steamed again in 1986, on the Great Central Line at Loughborough, its performance was transformed. Since then it has seen frequent main-line service, including the Settle to Carlisle route, producing outstanding performances.

'FLYING SCOTSMAN'

'Flying Scotsman' No. 4472, a class A3 Pacific 4–6–2, was built (originally as an A1) in 1923. Designed by Sir Nigel Gresley, it was constructed at the Doncaster works for the London and North Eastern Railway and used to advertise the 'Flying Scotsman' main-line service, helping to make it one of the best-known locomotives in Britain, if not the world. The 'Flying Scotsman' (by then renumbered 60103) ended service with British Railways in 1963.

The locomotive was one of the five Gresley Pacifics used on the London to Edinburgh route, using a modified tender and a water-trough system to allow a non-stop journey of 630 km (391 miles) in eight hours. After nationalization, the 'Flying Scotsman' ran on the Nottingham Victoria to London Marylebone service via Leicester Central, until 1963. It was first sold to Alan Pegler, a railway enthusiast and preservationist and had several subsequent owners, before being bought by the National Railway Museum in York in 2004. The 'Flying Scotsman' undertook tours to the USA in 1969 and Australia in 1988.

The 'Flying Scotsman' is now part of the National Collection in York. The National Railway Museum workshop team's full overhaul of the locomotive has involved complete dismantling, with all sections being measured with lasers to detect and correct any distortions. Restoration was delayed as more repair needs were discovered – perhaps to be expected after eighty years' hard service.

'KING EDWARD I'

'King Edward I' No. 6024, a Great Western Railway 'King' class four-cylinder 4–6–0 locomotive, was built at the Swindon works in 1930 and entered service in the same year. It was withdrawn from operations in 1962, along with all the other 'King' class locomotives. In 1974, 'King Edward I' was rescued from the Barry Scrapyard by the King Preservation Society (later renamed the 6024 Preservation Society Limited). After over fifteen years restoration, it was recommissioned in 1989.

Designed by Charles Collett and based on the earlier and successful 'Castle' class, the 'King' class locomotives were the heaviest and most powerful 4–6–0 engines ever to be built in Britain. Like the entire 'King' class, 'King Edward I' was modified during the 1950s. Improvements included a new type of boiler with higher superheat and better draughting using

double chimneys, giving sufficient additional power to cope easily with regular high-speed running and heavy loads.

'King Edward I' ran for thirty years on the Great Western Railway and Western Region of British Railways, frequently hauling prestigious passenger services such as the 'Cornish Riviera Express'. It continued to serve on major routes right up to its replacement by diesel locomotives in 1962.

The Preservation Society was determined to restore 'King Edward I' to main-line working and in 1990, it returned to hauling passenger trains on special steam excursions, heritage express routes, galas and charters. Further modifications to improve performance and safety have continued, including 'black-box' equipment for on-train monitoring and a water-wagon to allow long non-stop journeys.

'LOCOMOTION NO. 1'

A very early locomotive, 'Locomotion No. 1' hauled the first train that ran on the opening day of the Stockton and Darlington Railway in 1825. George Stephenson designed both the railway and the engine, but he had to persuade the directors of the Stockton and Darlington to try out steam power on the line, rather than use horses for haulage. A total of four locomotives were eventually built, at a price of £600 each.

The construction of 'Locomotion No. 1' owed much to the static beam-type pumping engines (some of which were designed by Stephenson) used to drain coal mines. It had two vertical cylinders set in the centre of the boiler and yokes attached to these turned the driving wheels, via a complicated arrangement of beams and connecting rods. The design did include one important technical feature: the locomotive linked its 0–4–0 wheel arrangement with coupling rods, rather than less robust chains or gears.

After being delivered by road from Newcastle, the engine was placed on the rails for its inaugural journey and it successfully hauled a train that included coal trucks and passenger wagons. During this first journey, it is thought to have reached a speed of 24 km (15 miles) per hour.

'Locomotion No. 1' experienced a broken wheel not long after it entered service and in 1828, the boiler exploded, killing the driver. After a thorough overhaul, it operated until 1841, when it was bought for use as a stationary pumping engine, having been made obsolete by more efficient locomotives, such as the 'Planet' class. 'Locomotion No. 1' became an early preserved locomotive in 1857, and is now part of the National Collection, on long-term loan to the Darlington Railway Centre and Museum.

'MALLARD'

'Mallard' No. 4468 is best known for holding the official world speed record for a steam locomotive in normal production. Built in 1938 for the London and North Eastern Railway, 'Mallard' was one of the A4 class Pacific 4–6–2 locomotives designed by Sir Nigel Gresley. It remained in service until 1963, after a working life covering almost 2.5 million km (1.5 million miles) and is now preserved at the National Railway Museum in York.

Gresley's design is famous for its streamlining, which he introduced after being impressed by the aerodynamic 'Flying Hamburger', a German diesel train working the Hamburg to Berlin route. 'Mallard' also had advanced mechanical design features. From the beginning, the A4 class utilized enlarged steam passages and a high boiler pressure. Later models (starting with 'Mallard') were also fitted with the 'Kylchap' double blastpipe and double chimney, to improve exhaust efficiency.

Gresley was keen to gain the speed record and the attempt was made on 3 July 1938, using 'Mallard' and a special test train: a 'dynamometer car' for taking accurate measurements and six other vehicles. The train started to the north of Grantham and headed south down the London and North Eastern Railway route. Descending Stoke Bank, south of Grantham, a maximum speed of 203 km (126 miles) per hour was recorded.

'Mallard' and its stable mates hauled express services until the end of the steam era and in 1961, 'Mallard' powered the last steam-hauled 'Elizabethan' London to Edinburgh express. The locomotive is now on static display in York, because of modern restrictions on steam traction, its speed record is unlikely ever to be broken.

'PUFFING BILLY' AND 'WYLAM DILLY'

'Puffing Billy', built in 1813–14, is probably the oldest surviving steam locomotive in the world. It also has a surviving sibling, 'Wylam Dilly' and for many years, experts were undecided which was the older, but their opinion now is that 'Wylam Dilly' has some more modern features incorporated, so it was possibly built afterwards. 'Puffing Billy' is preserved at the Science Museum, London and 'Wylam Dilly' is now part of the collection in the National Museum of Scotland in Edinburgh (formerly the Royal Museum).

'Puffing Billy' was designed for Wylam Colliery, near Newcastle upon Tyne, by the engineer, William Hedley, assisted by Timothy Hackworth and Jonathan Forster. Before it entered service, coal wagons were hauled along the track using teams of horses, but the Napoleonic wars had pushed up prices of fodder, making steam power a commercially-attractive alternative.

Before building the locomotive, Hedley carried out experiments to prove that heavy loads could be hauled using a smooth wheel and a smooth rail. He incorporated some technical improvements on earlier, more primitive steam engines, including vertical piston rods and a crankshaft. The original locomotive was too heavy for the cast-iron rails and broke them and had to be rebuilt with four extra wheels to distribute the weight more evenly.

The success of the final design enabled 'Puffing Billy' to run until 1862, when it was donated to the Science Museum – after almost fifty years' service.

'Wylam Dilly', possibly built a year later than 'Puffing Billy', took its name from a 'dilly', a term describing coal trucks used on the railway. The locomotive was used as the engine for a steam paddle boat before its eventual preservation in Edinburgh.

'ROCKET'

'Rocket' was designed by Robert Stephenson (1781–1848), a British engineer, assisted by his father, George. It was built for the Rainhill Trials, a competition held in 1829 to find the best type of locomotive for the new Liverpool and Manchester Railway. Stephenson's design was the clear winner, with a top speed of 47 km (29 miles) per hour.

'Rocket' was not the first steam locomotive ever built – primitive versions had been operating commercially since 1812 – but it combined a number of design features, which were very advanced for the time. Instead of a single flue carrying hot gases from the firebox to the boiler, 'Rocket' had twenty five copper tubes to heat the water: a 'multi-tube boiler'.

When the Liverpool and Manchester Railway was launched on 15 September 1830, there was an impressive opening ceremony, attended by the Prime Minister, the Duke of Wellington, with 'Rocket' attracting great interest. Sadly, William Huskisson, MP for Liverpool, was struck by 'Rocket' and killed: as his memorial states, 'striking terror into the hearts of the assembled thousands'.

The railway went on to become a great success. 'Rocket' is currently on display at the Science Museum, London. There is also a working replica at the National Railway Museum in York.

This was combined with a 'blastpipe', which used the steam exhaust to force air through the firebox. Stephenson also redesigned the cylinders and the connecting rods, which linked the pistons to the wheels. These features made 'Rocket' much more efficient than its rivals, and most steam locomotives built afterwards were based on Stephenson's design.

'STIRLING NO. 1'

The Great Northern Railway Stirling '8-ft Single' 4–2–2 is a striking locomotive, its design being dominated by the huge pair of 2.4 m (8 ft) driving wheels and domeless boiler. The first of the class, 'No.1' was built in 1870 and saw service until 1907. It is now preserved in the National Railway Museum, York.

Patrick Stirling, who joined the Great Northern Railway in 1866, designed his 'Singles' for speed; the large, efficient driving wheels were supplemented by outside cylinders, giving greater horse-power. Stability was enhanced by using a four-wheel leading bogie, instead of the normal two-wheeler. A total of fifty-three '8-ft Singles' were eventually built between 1870 and 1895 at the Great Northern Railway's Doncaster Works. During this 25-year production span, a number of modifications and improvements were introduced, including larger fireboxes and grates for some locomotives and increased boiler pressures.

The Singles soon became the railway's preferred locomotive for the prestigious London to York run. They worked a large part of the East Coast route and achieved impressive speed performances: engines of the class took part in the Railway Race to Edinburgh in 1888 and the Race to Aberdeen in 1895.

When 'No.1' was withdrawn from service in August 1907, it had completed almost 2.3 million km (1.4 million miles), a testament to the reliability of the design. In 1925, it took part in a tour of locomotives between Stockton and Darlington for the Railway Centenary celebrations. The preserved 'No. 1' at York is the only '8-ft Single' remaining – sadly, all others in the class have been scrapped.

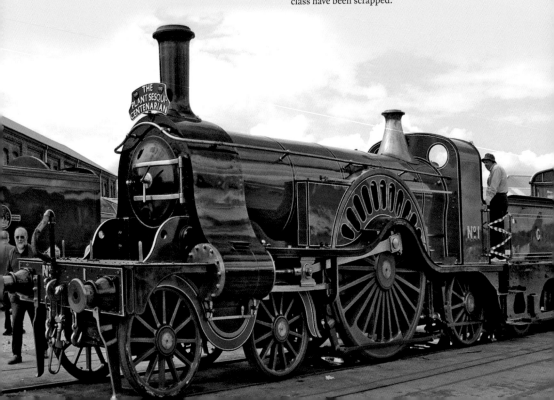

THOMAS THE TANK ENGINE

Thomas the Tank Engine first appeared in 1946 as one of the engines in a series of children's books written by the Reverend, Wilbert Awdry. Thomas soon became the star character, and the stories – never out of print – were made into a popular television show in the 1980s. The original illustrator, C. Reginald Dalby, based Thomas on an 0–6–0T class E2 shunting engine. The last five of these locomotives, introduced in 1915, had the characteristic extended water tanks which make Thomas so distinctive.

Many heritage lines operate 'Thomas the Tank Engine' specials for young railway enthusiasts – this Thomas is at Ropley station on the Watercress Line in Hampshire.

STEAM SPECIALS

Many of the great 'named trains' of the past have a fame that reaches far further than the railway enthusiast. Flagship expresses with name boards on their locomotives, such as the 'Flying Scotsman' and the 'Golden Arrow', conjure up a nostalgic vision of leather suitcases, attentive porters and record-breaking speeds. However, less romantic varieties also existed, such as the 'Blue Star Fish Special', a freight train that ran between London King's Cross and Aberdeen in the 1950s.

Quite how many Steam Specials have travelled our railways over the years is hard to estimate – there were certainly over 170 named services between 1915 and 1970. Many were relatively short lived, such as the 'Coronation Scot', a London, Midland and Scottish Railway passenger express, inaugurated in 1937 for the Coronation of King George VI. Like many services, it was discontinued at the outbreak of war in 1939. Other named trains reflected a different era, such as the boat train specials that linked with the great ocean-going passenger liners. These included 'The Cunarder', inaugurated in 1952, which ran from London Waterloo to Southampton. This service connected with the 'Queen Mary' and 'Queen Elizabeth' and operated until the late 1960s. Other boat trains called 'The Cunarder' operated on northern routes.

Although the first half of the twentieth century was the heyday of the 'named train', much earlier examples can be found. One of these was the Irish Mail, which first ran from Euston to Holyhead in 1848.

In addition to the named diesel or electric powered special trains which still run, many steam-hauled special trains now operate excursion routes, often retaining historic names for their services – allowing an important part of railway history to live on.

GWR Castle Class 4–6–0 No. 5029 'Nunney Castle' hauling the
London Victoria to Worcester 'Cathedrals Express', July 2009.

'CAMBRIAN COAST EXPRESS'

The 'Cambrian Coast Express' was a Great Western Railway train that ran from London Paddington, via Aberystwyth to Pwllheli in North Wales, a distance of 375 km (230 miles). The line beyond Welshpool was operated by Cambrian Railways (who have given their name to the route) up to 1923. Typical of the smaller, pre-amalgamation railways, Cambrian valued its independence and Great Western Railway passengers normally had to change to a Cambrian Railways train at Shrewsbury. However, in 1921 the Great Western Railway reached agreement to run a daily through express from Paddington to Pwllheli.

The name Cambrian Coast Express only began to appear on official literature in 1927, by which time a much reduced summer service was running. By 1939, the through train ran on summer Saturdays only, taking 5 hours 35 minutes to reach Aberystwyth, using the 'Castle' class 4–6–0 locomotive for the first part of the journey. After Paddington, the next major station was Birmingham Snow Hill, where a lavish reconstruction had been completed in 1912, including waiting rooms with solid oak bars. Despite a huge public outcry, the old Snow Hill station was demolished in 1977.

The most picturesque part, of what was by then primarily a tourist route, was the stretch from Machynlleth to Pwllheli, via Porthmadog. This included a view of the Mawddach estuary and a crossing of Barmouth Bridge, with the mountains of Snowdonia in the distance.

Although the last named train on the route ran in 1991, parts of the original Cambrian Coast Express line have been followed by heritage steam trains, including a service from Machynlleth to Pwllheli in the peak summer season.

'CATHEDRALS EXPRESS'

The 'Cathedrals Express' was originally a British Railways train that ran between Paddington and Hereford (via Gloucester) from 1957 to 1965. A revived service in 1985 was diesel hauled and a steam-excursion service, covering parts of the old route, as well as additional locations, has operated subsequently.

Gloucester, one of the retained destinations, is typical of the route, with its dominant 900-year-old cathedral The railway development at Gloucester, originally built as the terminus of the Birmingham and Gloucester Railway in 1840, has a complex history, involving four separate railway companies, five separate stations and two gauges of track.

Main-line steam excursions on the new Cathedrals Express run to a variety of other destinations, including Bath and Bristol, Cambridge, Canterbury, Exeter, Oxford, Stratford upon Avon and York. Rolling stock includes a refurbished 1950s Pullman-style lounge car. A wide variety of preserved locomotives work the current route. These include the newly-built A1 'Tornado' No. 60163 Pacific, designed by Arthur Peppercorn and built in 1948–49 for British Railways. Unhappily, in the rush to remove steam power, all the highly successful 'Peppercorns' had been cut up by 1966. Not to be thwarted, a group of enthusiasts – The A1 Steam Locomotive Trust – built one from scratch, completing the job in 2008. Another locomotive used on the route is 'Sir Lamiel' No. 30777, a Southern Railway N15 'King Arthur' class 4–6–0, built in 1925 (now part of the National Railway Museum's collection) and still running regular passenger services.

'CORNISH RIVIERA EXPRESS'

In 1904 the Paddington to Penzance Great Western Railway service was rescheduled to complete its journey in 7 hours. Through trains from London had been running from 1867, but took several hours longer for the trip. To mark the new timings, the Railway Magazine organized a competition to name the service, which eventually became known officially as the 'Cornish Riviera', or 'Cornish Riviera Express'. Railway workers also referred to it as 'The Limited'. 'Slip coaches' were a feature of the service: carriages that could be dropped off on the move at various stations on the route – they were slowed by a guard using a handbrake. Boarding the correct coach at the outset was, therefore, important. Steam traction was replaced by diesel in 1958, and the London to Penzance route currently takes around five hours to cover the 490 km (305 mile) journey.

Highlights of the route include long stretches of scenic coastal railway in the south west.

Since the start of the named service in 1904, a variety of classic steam locomotives have worked on the route. The first trains were hauled by 'City' class 4–4–0 engines, replaced by the 'Star' class 4–6–0s in 1907. These were superseded by 'Castle' class 4–6–0s in 1924. In 1927, the recently introduced 'King' class 4–6–0s took over, but only as far as Saltash: they were too wide to cross Brunel's famous railway bridge into Cornwall. In the late 1950s, 'Princess Coronation' class 4–6–2s were among the locomotives working on the route.

'CUMBRIAN MOUNTAIN EXPRESS'

The 'Cumbrian Mountain Express' operates in the summer months, using steam locomotives to carry passengers over Shap Summit to Carlisle, returning via the Settle–Carlisle line.

The route includes picturesque and historically-important locations. Shap Summit, 4 km (2½ miles) south of the old Shap station, is the highest point on the West Coast mail line, at 280 m (914 ft) above sea level. The Settle–Carlisle line, one of the best known and most dramatic railway routes in the country, has the Yorkshire Three Peaks as a backdrop: Pen-y-ghent, Whernside and Ingleborough. The line includes the highest railway summit in England, Ais Gill, at 356 m (1169 ft). In 1913, just a short distance to the north of the summit, this was the scene of a tragic rail accident – one that would have been prevented by modern safety regulations. A warning signal was missed

and a train, which had stalled on the hill, was struck by another coming up behind, with 16 dead and 38 seriously injured. The route crosses the River Ribble via the well-known Ribblehead Viaduct. Designed by the engineer, John Crossley, and built between 1870 and 1874, the viaduct is 32 m (104 ft) high and spans 402 m (1320 ft). Because of its curvature, passengers have a good view of the structure as they cross over it. The long Blea Moor Tunnel is another interesting feature.

Locomotives currently in use on the Cumbrian Mountain Express include 'Princess Elizabeth' No. 6201, a 'Princess Royal' class Pacific 4–6–2 locomotive, built in 1933 at Crewe, the second of its type. 'Princess Elizabeth' was withdrawn by British Railways in 1962, but has regularly hauled trains as a preserved locomotive.

'THE JACOBITE'

The Fort William to Mallaig extension of Scotland's West Highland Railway line was built between 1897 and 1901. This is the route followed by the current 'Jacobite' steam passenger train. The 135 km (84 mile) round trip includes some impressive scenery. Starting from Fort William (near Ben Nevis, the highest mountain in Britain) the route passes close to the ruins of the thirteenth-century Inverlochy Castle. Between Corpach and Glenfinnan, the Jacobite crosses the 21-arch Glenfinnan viaduct, built out of concrete by Sir Robert McAlpine, between 1897 and 1901. After a halt at the village of Glenfinnan, the route passes Loch Morar, considered to be Britain's deepest fresh-water loch. At the next stop at Arisaig, the most westerly railway station in Britain, the 'Small Islands'

of Rhum, Muck, Eigg and Canna are visible on a clear day. The terminus at Mallaig is next to Loch Nevis, Europe's deepest seawater loch. For those passengers not making the return journey, there is the option of a visit to the Isle of Skye, via the Caledonian MacBrayne ferry.

The line is well known for the operation of 'Olton Hall' No. 5972, a 4–6–2 Great Western Railway 'Hall' class steam locomotive, which has featured in the *Harry Potter* series of films. Built in 1937 at Swindon, it first saw service in South Wales, and was withdrawn in 1963. 'Olton Hall' operates with a matching set of coaches.

'ORIENT EXPRESS'

The story of the 'Orient Express' is a complex one. There has never been a monopoly on this famous name – by the 1920s and 30s, there was a whole interconnecting network of 'wagons-lits' (sleeping cars) company trains, called (at least as part of their name) 'Orient Express', in addition to the original Orient Express itself. For example, Agatha Christie's *Murder on the Orient Express* is set on the 'Simplon Orient Express', not the original.

The original 'Orient Express' is best seen as the 'first' one, rather than the 'real' one, for the reasons given above. As the 'Express d'Orient', it first left Paris in October 1883, a long-distance passenger train, originally operated by the Compagnie Internationale des Wagons-Lits. In its prime in the 1930s, it ran daily from Paris to Istanbul, taking three nights for the journey. Steam locomotives used for the service included the powerful Chapelon 4–6–2 Pacifics. The successor of this service (much truncated) made its last trip in December 2009, by which time it had evolved into an Austrian Railways train.

The 'Orient Express' name still lives on, including a luxury special train that travels between London, Venice, Istanbul and other continental destinations using restored vintage sleeping cars. There are also 'Orient Express' day trips in Britain. Many of these excursions are steam-hauled, using vintage Pullman-type carriages from the 1930s. Steam locomotives used on the British routes include 'Tangmere' No. 34067, an ex-Southern Railway 'Battle of Britain' class 4–6–2 known collectively as 'Light Pacifics'.

'SCARBOROUGH SPA EXPRESS'

The 'Scarborough Spa Express' running today has its origins in a named train operated by the London and North Eastern Railway Company. In 1927, the company introduced the 'Scarborough Flier'. An express service ran from London to York, then the locomotive was changed and the journey continued to Scarborough and Whitby. The war put a stop to the service in 1939, but in 1950, it was introduced as a summer-only route, renamed (or rather re-spelt) the 'Scarborough Flyer'. This ended in 1963. A new Scarborough Spa Express began operations in 1981, to mark the re-opening of the town's rebuilt Spa Complex. The original locomotive turntable pit, built in 1890, had to be excavated to make way for a new one, at considerable expense.

The current Scarborough Spa Express route, steam hauled to date, leaves from York. It then follows a circular route through the North Yorkshire countryside, calling at Knaresborough, Harrogate and Leeds before returning to York and then travelling on to Scarborough. Features of interest on this scenic route include Knaresborough Viaduct, which crosses the River Nidd. Construction started in 1846, but its opening was delayed until 1851, as the four-span stone bridge collapsed just before completion.

The numerous steam locomotives that have been used on the route since 1981 include 'Leander' No. 5690, a London, Midland and Scottish Railway 'Jubilee' class 4–6–0, built in 1936 at Crewe, and withdrawn in 1964. Like many preserved locomotives, 'Leander' was rescued from the Woodham Brothers' scrapyard in Barry.

'SHAKESPEARE EXPRESS'

The 'Shakespeare Express' is a steam-hauled service which runs from Birmingham Snow Hill and Birmingham Moor Street, via Tyseley and Henley-in-Arden to Stratford-upon-Avon. The return journey is often on a more easterly route, via Claverton and Lapworth, where the top permitted speed of 96 km (60 miles) per hour can be reached. The journey time is around one hour each way.

The route takes in Birmingham's urban and historic landscape, as well as attractive Warwickshire countryside, passing through working farms and small villages. Features visible at the start of the journey include the Bullring and Rotunda next to Moor Street Station; Birmingham City Football Club, seen from Bordesley station; and the Small Heath Mosque. The Tyseley Locomotive Works, which maintains a range of steam locomotives, is open to the public for a number of days each year. Further down the line, the route crosses the Stratford-upon-Avon Canal just before Whitlocks End; closer to Stratford, the Edstone Aqueduct can be seen.

The regular steam locomotive used for the service is 'Rood Ashton Hall' No. 4965, a 4–6–0 'Hall' class built for the Great Western Railway in 1929 at Swindon for mixed traffic duties. It was able to haul freight and express passenger trains and now carries authentic GWR livery. Other steam locomotives used to work the line include 'Earl of Mount Edgcumbe' No. 5043, a Great Western Railway 'Castle' class built in 1936. Rolling stock incorporates restored ex-British Railways Mk 1 Pullman dining cars, first introduced in the 1960s.

'WELSH MARCHES EXPRESS'

'Welsh Marches', an historical term with origins in the Middle Ages, today does not refer to an exact geographical area, but to the border region between England and Wales – from Chester in the north to Newport and Gloucester in the south. The Welsh Marches rail route, which can vary according to the sections and stations chosen, has always been popular for steam excursions. Scenic stretches include parts of rural Shropshire and the edges of the Brecon Beacons National Park at Abergavenny. Hereford has the most interesting station on the route. Constructed in 1853 for the Shrewsbury and Hereford Railway and designed by R. E. Johnson in mid-Victorian Gothic, the building is well preserved with impressive lancet windows.

Regular rail tours on the route (initially between Shrewsbury and Newport) were organized in the 1980s. These 1980s trains were hauled by a variety of preserved locomotives, including 'King George V' No. 6000, an ex-Great Western Railway 'King' class 4–6–0 built in 1927.

More recently, tours along the route have run from Shrewsbury to Worcester, via Hereford, Abergavenny and Gloucester. Two Great Western Railway 'Castle' class 4–6–0 locomotives were used for haulage: 'Earl of Mount Edgcumbe' No. 5043 (first named 'Barbury Castle'), built in 1936; and 'Nunney Castle' No. 5029, built in 1934. Both were originally withdrawn from service in 1963.

A line-up of locomotives at the Didcot Railway Centre, including GWR 'Earl' class No. 5051 'Earl Bathurst' and 'Castle' class No. 5029 'Nunney Castle', February 2010.

STEAM CENTRES, MUSEUMS AND WORKS

Public interest in the steam railway era is reflected in the many museums, steam centres and other locations where historic railway memorabilia are gathered together, ranging from collections of locomotives and other rolling stock to tickets, posters and timetables from our railway heritage.

Some of the larger railway collections are associated with the 'railway towns' that grew into focal points for the network in the 19th century. The National Railway Museum at York is one example. The coming of the railway in the late 1830s revolutionized life in the city. By the 1850s, a dozen trains a day ran between London and York, carrying over 340,000 passengers a year. A railway museum opened there in 1928, dominated at first by items from the London and North Eastern Railway. The Steam Museum of the Great Western Railway at Swindon and the Crewe Heritage Centre (formerly known as 'The Railway Age') are other examples of railway town setting. The Swindon museum includes a celebration of the achievements of the hundreds of thousands of 'navvies' (a word first used for the canal workers or 'navigators') who were employed in a vast amount of tunnelling, embankment building and bridge construction in the nineteenth century.

Other Steam Centres have a more local emphasis. One example of these is the East Anglia Railway Museum, near Colchester, in Essex. Like many such centres and museums it has been granted charitable status and relies heavily on volunteers. The locomotive collection includes an 0–6–2 side tank engine No. 7999, built in 1924 to the original A. J. Hill 1914 Great Eastern Railway design. This was used to haul suburban trains out of Liverpool Street and Fenchurch Street Station and since preservation has been seen regularly on the North Norfolk Railway.

BARRY SCRAPYARD

The story of the Barry Scrapyard, famous as a source of preserved steam locomotives, has its origins in 1955 with the British Railways Modernization Plan, which involved a rapid phasing-out of steam. Many steam locomotives were scrapped after only a few years in service. The result was a large quantity of surplus engines and when the numbers overwhelmed British Railways' ability to scrap them, a private company was contracted to help: Woodham Brothers Ltd based in Barry, South Wales. After the owner, Dai Woodham, won the contract in 1958, he had to go to the Swindon Works for a week to learn the complex business of scrapping steam locomotives – a daunting task for a traditional scrapyard.

The yard subsequently received large numbers of wagons and other rolling stock as well as locomotives – by 1968, 297 engines had been sent for scrap. Fortunately for steam preservationists, it made commercial sense for Woodham Brothers to start with the easier task of cutting up wagons, leaving many of the locomotives in store.

This presented a golden opportunity for rail enthusiasts, who raised money to buy and restore locomotives. Even at 'scrap prices', these were expensive items and Dai Woodham helped greatly by often allowing payment by instalments. The many preserved locomotives rescued from the Barry Scrapyard include the ex-London, Midland and Scottish 'Leander' No. 5690, a 'Jubilee' class 4–6–0 rescued in 1972.

Following Dai Woodham's retirement, the remaining ten locomotives were taken into store by the Vale of Glamorgan Council. The last to leave the Barry Scrapyard, in 1990, was a Great Western Railway '4575' class ('Small Prairie') 2–6–2T, No. 5553.

BARROW HILL RAILWAY CENTRE

Construction of the steam roundhouse at Barrow Hill, near Chesterfield in Derbyshire, was completed in 1870. This historic circular building, with a turntable in the centre, was used for storing and repairing locomotives until it closed in 1991 – a working life of 121 years. It was at the centre of a complex of sheds and sidings. At its peak in the 1920s it was allocated ninety engines, reducing to around thirty when steam was phased out in 1965. Diesel locomotives then used the site until closure.

Before the site closed in 1989 the 'Barrow Hill Engine Shed Society' was set up to save the original roundhouse from demolition. After a process lasting almost ten years, which included getting the building Grade II listed and raising funds for its purchase and restoration, the engine shed was opened to the public in 1998. On the opening day, four steam engines were inside the shed, one of them a 'half-cab' Midland Railway '1377' class 0–6–0 goods engine, No. 41708, designed by Samuel Johnson and built in 1878. Most of these goods locomotives were built with open-backed cabs – hence the nickname. Withdrawal of this class began in 1928, although some remained in use until the 1960s. Other long-lived locomotives seen at the Centre over the years include a Great Eastern Railway 'Y14' class 0–6–0 No. 65462: this class was built between 1883 and 1913.

The Barrow Hill Railway Centre hosts regular galas of both diesel and steam locomotives, often joined by 'visiting' engines and rail tours off the main line.

DIDCOT RAILWAY CENTRE

The Didcot Railway Centre, in Oxfordshire, is based round the original Great Western Railway Engine Shed built in 1932. The Centre houses a large collection of steam locomotives, coaches, wagons, buildings and other historic items.

Much of the engine shed remains in its original condition, including smoke hoods running above the tracks, designed to carry smoke out through wooden chimneys. Nearby there is a hand-operated turntable, a replacement of the original. Other historic buildings include a World War II ash shed, used to prevent enemy aircraft spotting the glow from hot ashes when they were raked out of a locomotive and a railway policeman's hut.

The Didcot Railway Centre has preserved some of Brunel's broad gauge railway, with a width of 2140 mm (7 ft ¼ in).

This was finally abandoned in favour of today's 'standard' gauge in 1892. Much of the track at the Centre is 'mixed gauge', a complex arrangement with an extra rail between the two broad gauge lines.

The Centre houses a large collection of locomotives; there are usually some in steam as well as others on display or under restoration in the Locomotive Works. These include classic 'Hall', 'Manor', 'Castle' and 'King' class engines. The Didcot Railway Centre is also home to the 'Firefly', an accurate replica of the original 'Firefly' locomotive designed by Daniel Gooch in 1840 to run on Brunel's broad gauge Great Western Railway between London and Bristol. Some of the many coaches based at the Centre are regularly used on trains, with others on display. There is also a varied collection of over fifty railway wagons – mostly Great Western Railway stock.

THE MIDLAND RAILWAY MUSEUM, BUTTERLEY

The Midland Railway, Butterley, is near Ripley in Derbyshire and is based at Butterley station, Hammersmith station and Swanwick station, connected by a stretch of restored line.

The railway has its origins in a proposal by Derby Corporation in 1969 to create a working museum to celebrate the history of the original Midland Railway. A site was identified on a stretch of the old Pye Bridge to Ambergate line, part of which, west of the A4, had vanished due to road development. However, the remaining 6 km route (3½ miles) was available, even though the rails had been lifted. The Midland Railway Company Ltd (now called the Midland Railway Trust) was formed in 1973. After years of hard work by volunteers, the first passenger train ran over a short length of completed track in 1981.

The line has since been extended almost to Pye Bridge in the east and Hammersmith in the west. There is now a carriage and wagon department at Butterley and a museum and locomotive shed at Swanwick. Signal boxes from Kettering and other locations have been re-erected along the line. The restored line now crosses Butterley Reservoir on a large stone embankment which replaced an older bridge in the 1930s.

Locomotives that have visited the line include Furness Railway No. 20, the oldest working standard gauge locomotive in Britain. Built in 1863 by Sharp Stewart & Co of Manchester, it was one of a batch of 8 supplied between 1863 and 1866. Other preserved locomotives seen at the railway include 'Princess Elizabeth' No. 6201, a 4–6–2 Pacific built in 1933 for the London, Midland and Scottish Railway.

NATIONAL RAILWAY MUSEUM, YORK

The National Railway Museum was established after the Beeching Report recommended that British Railways should not be directly involved in museum operations and opened at Leeman Road in York in 1975. It was the direct successor to a number of museums previously run by British Railways (including the museum's predecessor in York). The new site in York is based on the site of a large steam locomotive depot and has been extended several times since 1975, including a viewing gallery over York station. There have also been major educational developments: the establishment of the Institute of Railways Studies, in partnership with York University; and the opening of the Yorkshire Rail Academy in 2004.

The collection includes posters, artwork, examples of signalling and telecommunications on the railways and a wide variety of other items, such as railway furniture, fixtures, fittings and timepieces. A comprehensive research archive includes documentaries and other film footage.

Taking pride of place in the museum is the large collection of over a 100 locomotives and almost 200 other items of rolling stock. Early examples of steam locomotives include 'Coppernob' No. 3, 0–4–0, built by Bury, Curtis & Kennedy in 1846 for the Furness Railway and withdrawn in 1900, after over 50 years of service. At the other end of the scale, the museum holds 'Evening Star' No. 92220, a powerful British Railways '9F' class 2–10–0, built at Swindon as late as 1960. Sadly, it proved to be not as long-lived as 'Coppernob', being withdrawn in 1965, as a result of the 'scrap steam' policy.

'A Railway Map' from
The Book of Trains
by Archibald Williams,
published by Thomas
Nelson and Sons in 1926
– an evocation of British
railways, which would
have been familiar to every
child at the time.

BIBLIOGRAPHY

Ahrons, E. L., The British Steam Locomotive, 1825–1925. London, 1927

Allan, Ian, Railway Colour Album. London, 1960

Allen, Cecil J. (revised by B. K. Cooper), Titled Trains of Great Britain, Littlehampton, 1983

Atterbury, Paul, Branch Line Britain, Newton Abbot, 2006

Atterbury, Paul, Along Lost Lines, Newton Abbot, 2007

Breckon, Don, The Railway Paintings of Don Breckon. London, 1982

Burton, Anthony, and Scott-Morgan, John, Britain's Light Railways. Ashbourne, 1985

Carter, Ernest F., The Observer's Book of Railway Locomotives of Britain. London, 1955

Chakra, Narisa, Terence Cuneo, Railway Painter of the Century. London, 1990

Clay, John F. (ed), Essays in Steam. Shepperton, 1970

Cole, Beverly, and Durack, Richard, Happy as a Sand Boy: Early Railway Posters. York, 1990

Cole, Beverly, and Durack, Richard, Railway Posters, 1923–1947. York, 1992

Cooper, B. K., Great Western Railway Handbook. London, 1986

Earnshaw, Alan, An Illustrated History of Trains in Trouble, 1868–1968. London, 1996

Ellis, C. H., Some Classic Locomotives. London, 1949

Ellis, C. H., British Railway History. London, 1954–59

Ellis, C. H., Railway Art. London, 1977

Fearnley, Alan, The Railway Paintings of Alan Fearnley. London, 1987

Ferneyhough, Frank, The History of Railways in Britain. London, 1975

Freeman Allen, Geoffrey, Railways in Britain. London, 1979

Garratt, Colin, and Wade-Matthews, Max, The Illustrated Book of Steam and Rail. London, 2001

Gordon, W. J., Our Home Railways: How They Began and How They are Worked. London and New York, 1910

Grinling, C. H., The Great Northern Railway. London, 1898

Herring, Peter, Classic British Steam Locomotives. Leicester, 2000

Holland, Julian, The Lost Joy of Railways, Newton Abbot, 2009

Hollingsworth, Brian, The Guild of Railway Artists: The Great Western Collection. Dorset, 1985

Johnson, Peter, An Illustrated History of the Welsh Highland Railway, Yeovil, 2002

Jordan, Owen, Jordan's Guide to British Steam Locomotives. Rotherham, 2003

Kennedy, Ian, and Treuherz, Julian, The Railway: Art in the Age of Steam. London and Kansas City, 2008

Lloyd, Roger, Railwayman's Gallery. London, 1953

Moore, F, British Expresses, 1898. London, 1898

Nock, O. S., The British Steam Railway Locomotive, 1925–1965. London, 1966

Nock, O. S., Steam Railways of Britain. London, 1967

Nock, O. S., British Steam Locomotives at Work. London, 1967

Nock, O. S., Steam Locomotive (2nd ed). London, 1968

Nock, O. S., Railways at the Turn of the Century, 1895–1905. London, 1969

Nock, O. S., Railways at the Zenith of Steam, 1920–40. London, 1970

Nock, O. S., Rail, Steam and Speed. London, 1970

Nock, O. S., Railways in the Years of Pre-Eminence, 1905–19. London, 1971

Nock, O. S., The Dawn of World Railways, 1800–1850. London, 1972

Nock, O. S., Railways in the Formative Years, 1951–1895. London, 1973

Palin, Michael, Happy Holidays, The Golden Age of Railway Posters, London, 1987

Reed, Brian, Modern Locomotive Classes. London, 1945

Ross, David, The Illustrated History of British Steam Railways. Bath, 2004

Richard, J., and MacKenzie, J. M., The Railway Station: A Social History. Oxford, 1986

Richards, Jon, Cutaway Trains. London, 1998

Rolt, L. T. C., Red for Danger. Newton Abbott, 1966

Simmons, Jack, The Railway in Town and Country, 1830–1914. Newton Abbott, 2000

Simmons, Jack, The Railways of Britain. London, 1986

Simmons, Jack, The Victorian Railway, London, 2009

Spence, Jeoffry. Victorian and Edwardian Railway Travel from Old photographs. London, 1977

Tomlinson, W. W., The North Eastern Railway, its Rise and Development, London, 1914

Tuplin, W. A., British Steam Since 1900. Newton Abbott, 1969

Unwin, Philip, Travelling by Train in the Edwardian Age. London, 1979

Vaughan, Adrian, Tracks to Disaster. Hersham, 2000

Westwood, J., Railways at War. London, 1980

Williams, Archibald, The Book of Trains. Great Britain, 1926

Wolmar, Christian, Fire and Steam, London, 2008

INDEX

About the Authors

Jon Mountfort is an engineer and writer who has studied the origins and application of the steam engine for over thirty years.

Tom Dodds has written on road and rail transport history, and regional and international transport issues.

Tony Evans is a writer and a member of the Wensleydale Railway Association, and he has a particular interest in pre-1900 railways.

David Adams is a long-time railway and history enthusiast who has written many articles on subjects including steam railways, canals and historical buildings.